Developed and Published by

AIMS Education Foundation

This book contains materials developed by the AIMS Education Foundation. **AIMS** (**A**ctivities **I**ntegrating **M**athematics and **S**cience) began in 1981 with a grant from the National Science Foundation. The non-profit AIMS Education Foundation publishes hands-on instructional materials that build conceptual understanding. The foundation also sponsors a national program of professional development through which educators may gain expertise in teaching math and science.

AIMS Education Foundation
P.O. Box 8120, Fresno, CA 93747-8120 • 888.733.2467 • aimsedu.org

ISBN 978-1-60519-032-7

Printed in the United States of America

I Hear and I Forget,
I See and I Remember,
I Do and I Understand.

-Chinese Proverb

Sense-able Science

Table of Contents

Sense-able Science

Our Five Senses

People use their senses to explore and describe their surroundings and themselves.

Most often, senses are used in conjunction with each other.

Sensory perceptions can stimulate emotions and feelings.

Different senses provide different information.

Taste

Humans use their sense of taste to distinguish between flavors, including sweet, sour, salty, and bitter.

The sense of taste is interrelated with the senses of smell and sight.

Sight

Sight is dependent on the availability of light.

When sight is absent, a human depends on other senses for information.

Hearing

Our sense of hearing enables us to identify the sounds of familiar objects and events.

Smell

Humans find smells to be distinctive, identifiable, and memorable.

The sense of smell can be desensitized.

Touch

The sense of touch enables us to discriminate textures, shapes, and sizes.

Sense-able Science

An Overview of the Five Senses

Are you ever amazed when you consider how we function day by day? We interact with our environment and control our own bodies. We do this by performing three operations: 1) we perceive our world with our senses, 2) we receive and process the information from our senses through our nervous system, 3) we react to that processed information with our muscles and skeletal system. To simplify, we smell the chicken grilling on the barbecue, our brain tells us that our stomach is not full, and we fill our plate to eat—we sense, we process, and we react!

In the primary grades, students learn a simplified classification scheme of the human's five main senses—touch, taste, smell, seeing, and hearing—that provide them with information about their world. Each of these five senses has a sensory organ associated with it: The organ for touch is the skin, the organ for taste is the tongue, the organ for smell is the nose, the organ for seeing is the eye, and the organ for hearing is the ear. These organs contain receptors that receive information and relay it as electrical signals called nerve impulses to the brain.

Nerves carry the impulses of information from the sensory organs to the brain. The thalamus, a small region in the center of the brain, is the first stop for many impulses. It helps to sort out, interpret, and compare the information from the different sensory organs. The thalamus acts like a post office, which sorts the letters it receives and sends them to the correct addresses.

Information is sent from the thalamus to various sensory centers of the cerebral cortex. The cortex is a very thin layer that covers the top and sides of the brain. Impulses from each sense are localized in discrete regions of the cortex where their information is received and processed. Severe brain damage may cause particular sensory losses, such as blindness or deafness. Since brain cells cannot normally reproduce, once a region of the brain is destroyed, such sensory losses are often permanent. In some cases, however, training can cause other undamaged regions of the cortex to take over some of the lost functions.

Sight

Have you ever watched a bird soar in the sky or the spray created behind a moving truck after a rain? Have you ever looked closely at the petals of a rose? Sights such as these are common enough that often we hardly pay any attention to them, yet the ability to see sights such as these is a marvel in itself. Vision allows us to view the light from stars that are billions of miles away and to thread a needle. Our sense of sight profoundly affects our lives. Eighty percent of the information received by the brain comes in through our eyes. Our sense of sight enables us to know the size and shape of objects, how near they are, and how fast they are going.

Our sense of sight is the most highly developed of the five senses studied in this book. Light enters the eye through an opening called the pupil. It is focused by a lens and projected onto the retina at the back of the eyeball. The retina contains the light-sensitive receptors called rods and cones that convert the light into nerve impulses. The nerve impulses are sent along the optic nerve to the brain.

The vision center of the brain is located at the back of the cerebral cortex. Information from the receptors is sifted, coordinated and interpreted here.

I Can't Imagine

I can't imagine being blind
What things I couldn't do or see
But I think I know someone who'll
Explain it all to me.

I know a man who has a dog
To guide him where he needs to go,
Who sees the steps and holes and curbs
And tells his master so.

Ruff goes along to restaurants,
Rides buses, goes to school.
When it's not safe to cross a street,
He's stubborn like a mule.

My friend can use his fingertips
To read his favorite books;
Small "letter bumps" form words in braille
They're handy when he cooks.

The braille knobs on his kitchen stove,
The braille watch on his wrist,
Help him to do what I can do
Without sight, there's the twist.

Another thing he does that's neat
To know how much he spends,
Is fold his money different ways
To tell the fives from tens.

By counting steps around his house,
He knows how it's arranged;
And he can find things easily
As long as nothing's changed.

There's one thing he finds funny, too,
How some folks speaking to him yell.
They seem to think, that since he's blind,
His hearing's gone as well!

So, when I think about my friend,
I'm grateful I can hear and see.
But now I know if I could not,
I'd learn to do things differently.

Brenda Dahl

My Eyes Can See

My eyes are brown
Or blue or gray
Or black or green;
They're made that way.

I look at you,
You look at me,
Our eyes are made
Such a wonderful way
So we can see!

My eyes need light
So I can see
Such colors bright
Are there for me.

Just look around
And you'll agree,
The world is full
Of such beautiful things.
Our eyes can see!

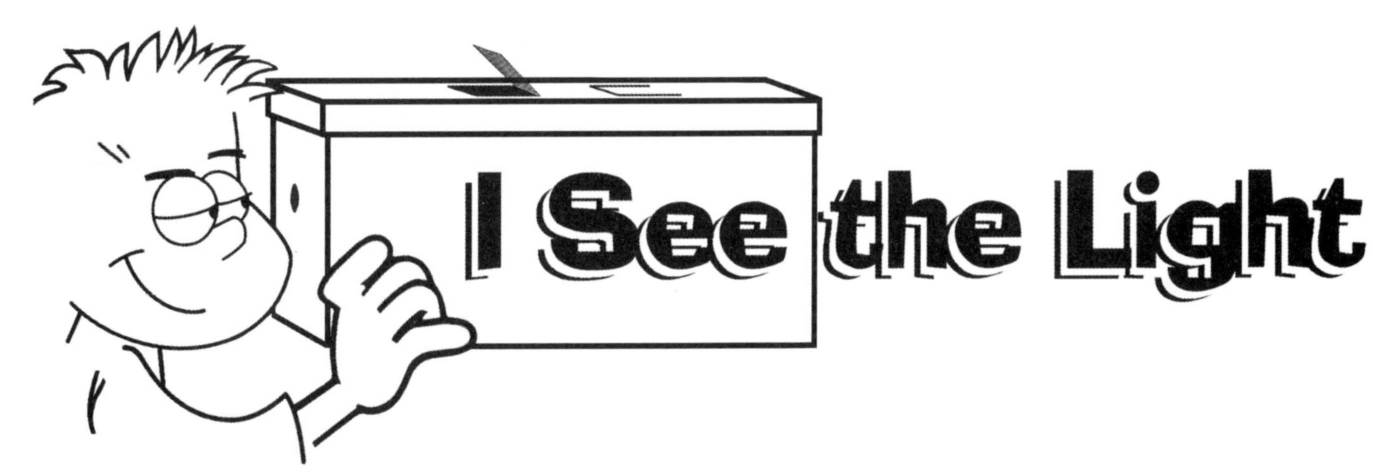

I See the Light

Topic
Sense of sight

Key Question
What do your eyes need in order to see?

Learning Goal
Students will discover that our eyes need light for them to see objects.

Guiding Document
Project 2061 Benchmark
- *People use their senses to find out about their surroundings and themselves. Different senses give different information. Sometimes a person can get different information about the same thing by moving closer to it or further away from it.*

Science
Life science
  human body
    senses

Integrated Processes
Observing
Comparing and contrasting
Recording data
Interpreting data
Applying

Materials
For the class:
  eight boxes
  teacher-made dioramas (see *Management 2)*

For each student:
  student pages
  crayons

Background Information
Our eyes contain photoreceptors, specialized sensory cells that detect light and differences in light intensity. These sensory cells are called rods and cones. Rods are much more light-sensitive than cones. Vision in dim light, such as moonlight, is almost entirely due to reception by the rods. There are three different types of cones, each one sensitive to a different color of light; together they provide color vision. With no light, we are unable to see. With dim light, colors are difficult to detect and we see mostly in black and white.

Management
1. Boxes that can be completely closed, such as shoe boxes or paper boxes, are necessary for this activity.
2. Glue or tape objects into each of eight boxes to make a diorama.

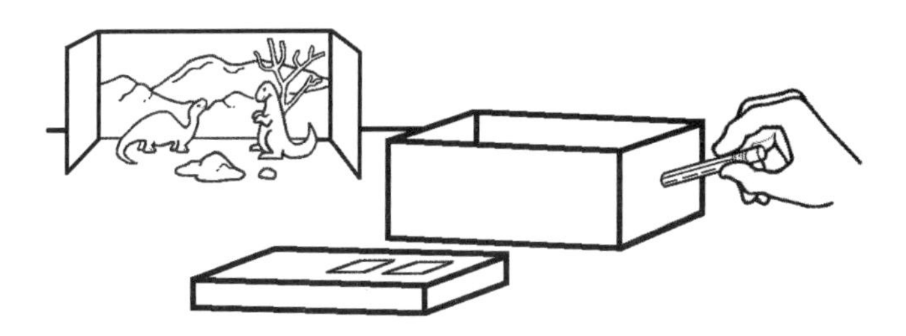

3. Use a pencil to poke a small hole at one end of the box for a "peep" hole.
4. Cut two flaps on the top of the box that can be opened and closed. The size of the flap depends on their placement and the size of the box. Try to place the flaps near the middle of the top of the box.

Procedure
1. Ask your students what they use to see.
2. Tell them that you are going to pass some boxes around and they are to peek into each box. (The flaps should be closed.) They are then to draw a picture of what they see on their recording pages in the space marked *First I See.* (This will usually be a black drawing.)
3. After students have peeked into the boxes and made their responses on their pages, demonstrate how to open one of the top flaps on their boxes.

4. Tell your students to pass the boxes around their group and to peek into the box with one flap open and to record their answers in the *Second I See* space. (Students should be able to see silhouettes.)
5. Once all the students have recorded their observations, tell them to open the second flap on the top of the box and to once again pass the box around. Instruct the students to record their observations in the *Third I See* space. (Students should be able to see definite pictures with color.)
6. Finally, allow the students to draw a design of a peep box they would like to make to share with other students.

Connecting Learning

1. Was your first drawing the same as your second?
2. Why do you have different drawings when you looked at the same thing three times?
3. What was different about the box the second time you looked into it?
4. What did your eyes need in order to see the objects in the box?
5. When could you see color: The first, second, or third time you looked into the box?
6. What did your eyes need more of to see the color? [light]
7. What are you wondering now?

Extensions

1. Have groups exchange boxes and repeat the activity.
2. Allow the students to design and build their own peep boxes and record what they can see in them with different amounts of light. Allow them to pass the boxes around to other groups. Have the students record what they see with *no light*, *a little light*, and *more light*.
3. Find a room at school (bathroom, teacher supply room, a room on stage, etc.) that can be made very dark. Take the students, in pairs, into the room and turn a light on. Ask them to count how many fingers their partner is holding up. Turn out the lights and again ask them to determine the number of fingers their partners are holding up. Again, they will see that light is needed for them to even see their partner next to them.

Home Links

1. Send the assignment of designing and building their own peep boxes home and make it a family project. Ask the students to bring their creations back to school to share with others.
2. Instruct the students to try to see in their bedrooms at night when all the lights are out. If they can still see, ask them to find out where the light is coming from that is helping them to see (outside street lights, lights in another room, the hall light, a night light, the moon, etc).

I See the Light

First I see

Second I see

I See the Light

Third I see

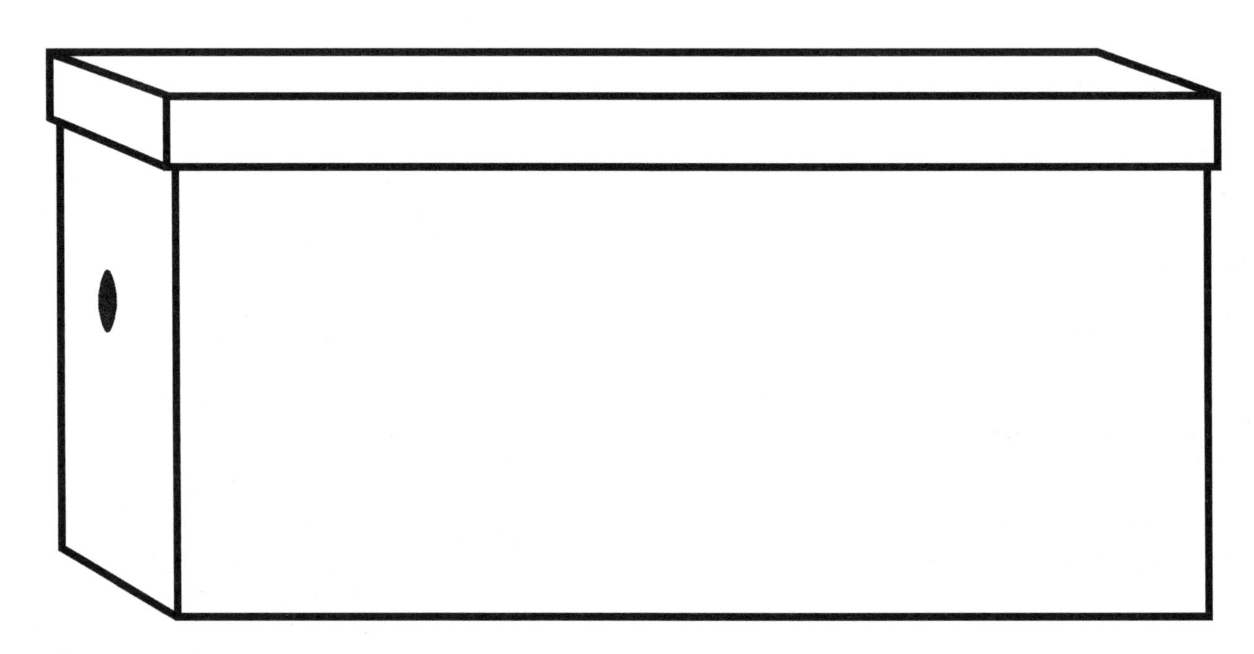

A plan for my own peep box

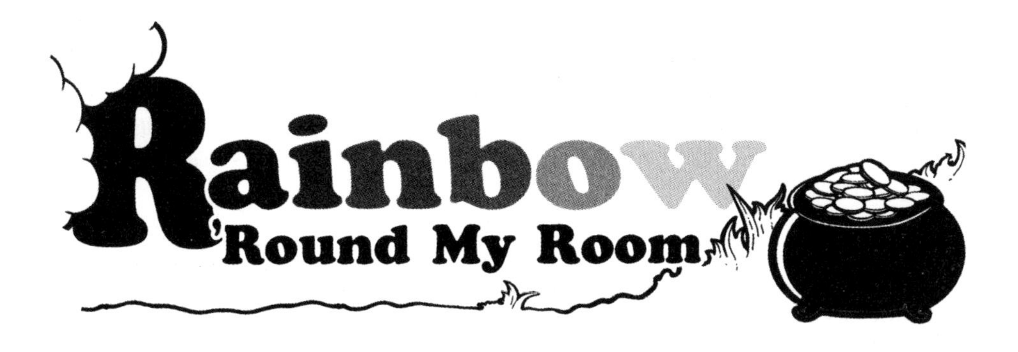

Topic
Sense of sight

Key Question
How many rainbow colors can we find in our room?

Learning Goal
Students will become aware that they use their sense of sight to classify colored objects.

Guiding Documents
Project 2061 Benchmark
- *People use their senses to find out about their surroundings and themselves. Different senses give different information. Sometimes a person can get different information about the same thing by moving closer to it or further away from it.*

*NCTM Standards 2000**
- *Sort and classify objects according to their attributes and organize data about the objects*
- *Represent data using concrete objects, pictures, and graphs*
- *Count with understanding and recognize "how many" in sets of objects*

Math
Charting
Counting

Science
Life science
  human body
    senses

Integrated Processes
Observing
Comparing and contrasting
Sorting and classifying
Collecting and recording data
Interpreting data
Communicating

Materials
Color Chart, enlarged
I Can See a Rainbow big book (see *Management 2)*
Colored markers and/or colored pencils
Rainbow Song
Construction paper: red, orange, yellow, green, blue, indigo, violet

Background Information
Teaching the recognition of color is one of the first concepts taught in primary classrooms. Tying color lessons to eyesight seems natural for students. You may find that some students may have difficulty distinguishing between certain colors. For example, shades of red may look orangish to some. Some of the students, especially the boys, may be colorblind or color deficient. Be sensitive to these differences and use the opportunity to explain that we are all unique individuals. Going into the scientific information regarding the reasons for these differences would not be developmentally appropriate at this age. Simply learning that there are differences, or experiencing the differences, is enough.

Management
1. Enlarge the *Color Chart* page to fill a sheet of chart paper.
2. Construct a class *I Can See a Rainbow* big book (see instructions). Do not put any objects in the pockets at this time.

Procedure
1. Sing the *Rainbow Song* while flipping through the big book you have made. As you sing the song, flip through to the next color page of the book.
2. Hold up a piece of red construction paper and say the color. Have students look around the room until they find objects that are the same color as the paper. Ask them to name those objects.
3. As red objects are named, invite a student to draw these items in the appropriate place on the *Color Chart* using a red marker. Once several items have been recognized, proceed in the same manner with the next colors (orange, yellow, green, blue, indigo and violet) until you have finished all the colors in the *Rainbow Song.*

4. Tell the students that they are going to decide which items to put in the pockets of the class *I Can See a Rainbow* big book. Ask them what object they would like to use to represent red. Fill in the name of the object on the sentence strip at the bottom of the page. For example, I can see a red *apple*. Continue with the other colors.
5. After the students have dictated what to write in the blank spaces, ask them to draw or paint the objects to be put in the pockets. Encourage them to use different mediums such as chalk, marker, paint, or crayons.
6. Give the students an opportunity to make their own *I Can See a Rainbow* books. There are two different strips provided for the cover of the individual student's *I Can See a Rainbow* book. One has the title *I can see a rainbow*. The students then copy this sentence on the blanks below. The second title strip has *I can see a rainbow* ______________. The blank line is for the students to either dictate or write in the last part of the sentence. Examples might be: I can see a rainbow in the sky, or between the clouds, or over me. It is suggested that you use one or the other depending on the abilities of your students.

Connecting Learning

1. What colors did our eyes find in the classroom?
2. Let's count all the objects in each color and write the number in each section of our chart.
3. For which color did we have the most objects recorded? How could you tell? How many more _______objects did we see than _______ objects?
4. Which color has the least number of objects recorded?
5. If you close your eyes, can you still see all the colors in our room? What do we need to see things?
6. Did your eyes find colors that are not on the graph? What are they? How can we show them on a chart or graph?
7. What are you wondering now?

Extensions

1. Have the students bring old magazines from home and cut out pictures of objects. Specify one color each day. Have them glue their pictures to a piece of the same color construction paper. When you are finished with the various colors, staple them all together so each child can have a personal color book.
2. Show the students a prism and point out how a rainbow can be made by shining light through the prism.
3. Talk about how different colors make you feel. What do you think of when you hear about the color green? ...red? ...blue?, etc.

Curriculum Correlation

Language Arts

Have students finish the sentence: "Today I found a/an (color word)(object) in our classroom." Have them select the color words from a list of colors put on word strips.

Literature

Carle, Eric. *Brown Bear, Brown Bear, What Do You See?* Henry Holt and Company. New York. 1996.

Cottin, Menena. *The Black Books of Colors*. Groundwood Books. Toronto, CA. 2008.

Dodd, Emma. *Dog's Colorful Day: A Messy Story About Colors and Counting*. Penguin Puffin Books for Young Readers. New York. 2003.

Ehlert, Lois. *Planting a Rainbow*. Voyager Books. New York. 1992.

Freeman, Don. *A Rainbow of My Own*. Puffin Books. New York. 1978.

Jonas, Ann. *Color Dance*. Greenwillow Books. New York. 1989.

Lionni, Leo. *A Color of His Own*. Knopf Books for Young Readers. New York. 2006.

Peek, Merle. *Mary Wore Her Red Dress and Henry Wore His Green Sneakers*. Houghton Mifflin. New York. 1995.

Stinson, Kathy. *Red is Best*. Annick Press. New York. 2006.

Walsh, Ellen Stoll. *Mouse Paint*. Voyager Books. New York. 1995.

Home Links

1. Specify a different color each day. Give the students a piece of paper to take home to draw an object from their home of that color. Have students bring the paper back to school to display on the color wall.
2. Each day have students bring a small item (it must fit into a lunch sack) from home. Ask them to identify the color of the item and to try to find the written word that matches the color of the object.
3. Have students select a color that they should try to wear for the next day. If some students cannot wear the selected colors, have ribbons or pieces of construction paper in the selected color that can be pinned to their shirts.

Red, the top, the highest color,
Orange, we'll find it just below;
Yellow follows, bright and sunny;
Green, the middle one, we know.

Blue comes next, like sky in summer;
Indigo, is purple blue;
Violet, is the bottom color;
There's a rainbow just for you!

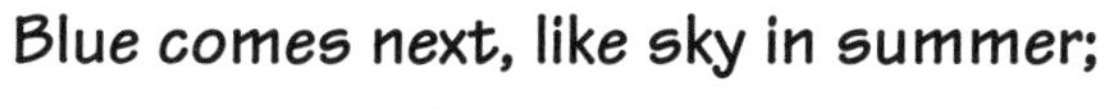

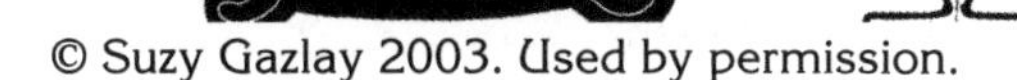

Rainbow 'Round My Room

I Can See a Rainbow—Big Book Instructions

1. Cut construction paper to the following sizes to make a big book with a white cover and pages the colors of the rainbow. Because the pages of the book have different dimensions, the book will have a layered look.

white	12" x 18" (cover)
red	12" x 12"
orange	12" x 13"
yellow	12" x 14"
green	12" x 15"
blue	12" x 16"
indigo	12" x 17"
violet	12" x 18" (back page)

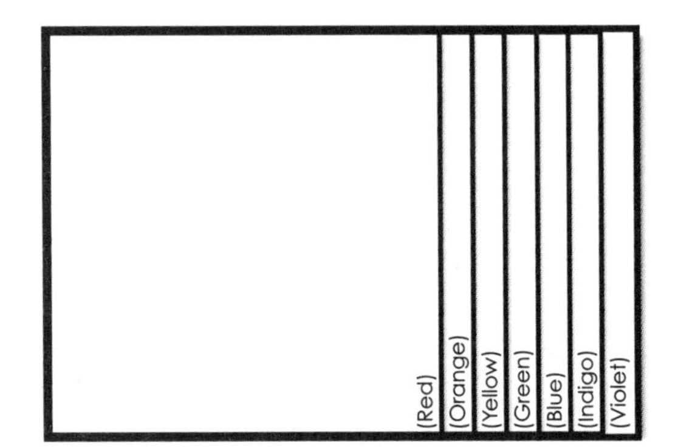

2. Fold a 3-inch strip up from the bottom of each page to form a pocket. To hold the pocket in place, staple twice along its right side.

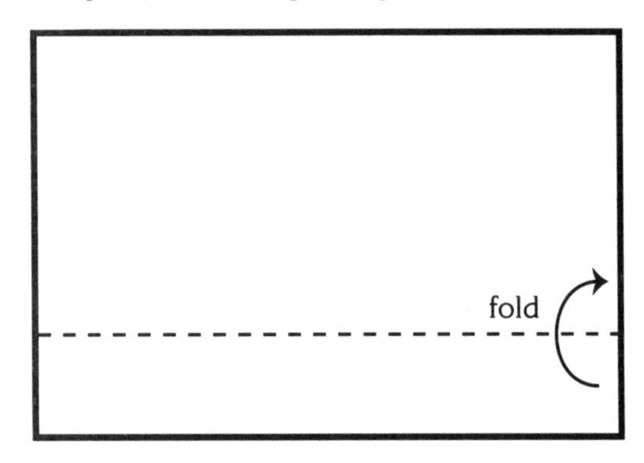

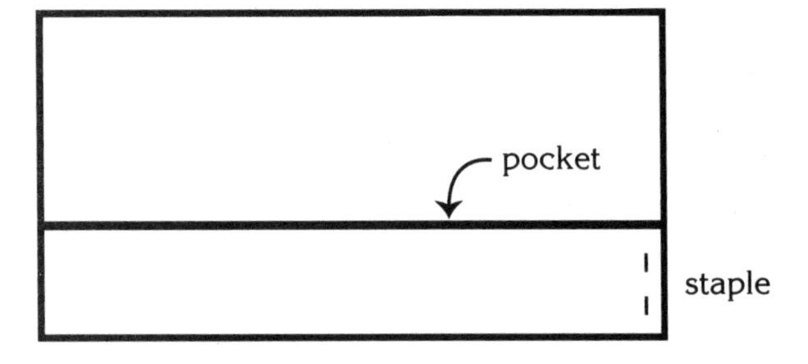

3. Assemble the pages in their proper order. Staple along the left side. Cut a 3" x 12" strip of any color to use for the binding on the left side of the book. Fold it in half as illustrated and glue it to the left side of the book.

4. Cut out the entire sentence strip if you want your students to copy the text. Cut out just the top script if you want them to read each strip and fill in their own objects. Glue the sentence strips onto each pocket.

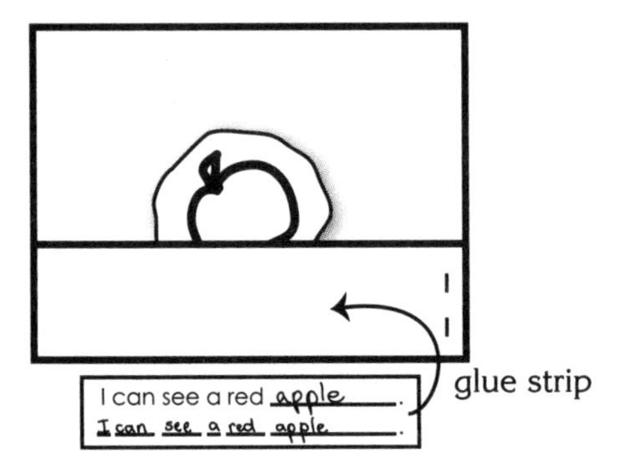

5. Allow students to illustrate different objects that would fit into the pockets. Encourage them to use a variety of mediums: chalk, paint, sponge painting, colored pencils, markers, etc. You may choose to use only one object for each pocket, or you can change the objects each time the book is read.

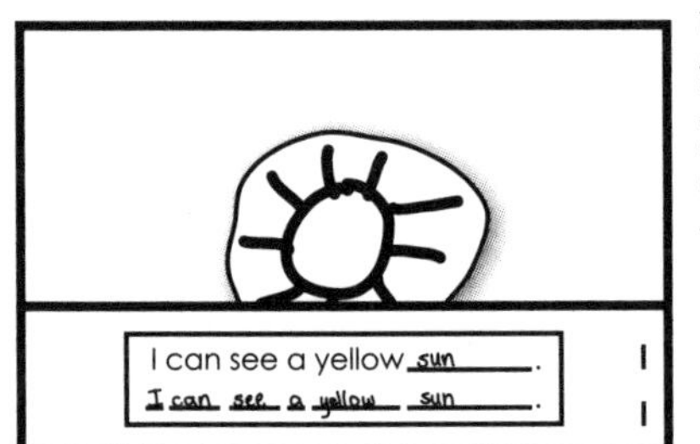

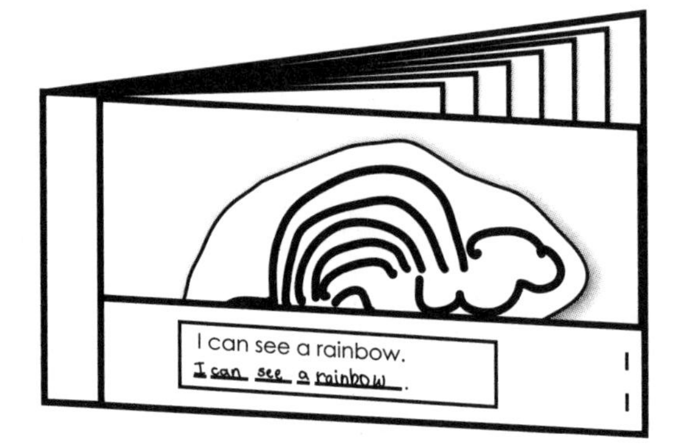

I can see a rainbow.

_ ___ ___ _ ________

I can see a rainbow ____________.

_ ___ ___ _ ________ ____________

I can see a red ____________.

_ ___ ___ _ ___ ____________

I can see an orange ______.

I can see a yellow ______.

I can see a green ______.

I can see a blue ______.

I can see an indigo ______.

I can see a violet ______.

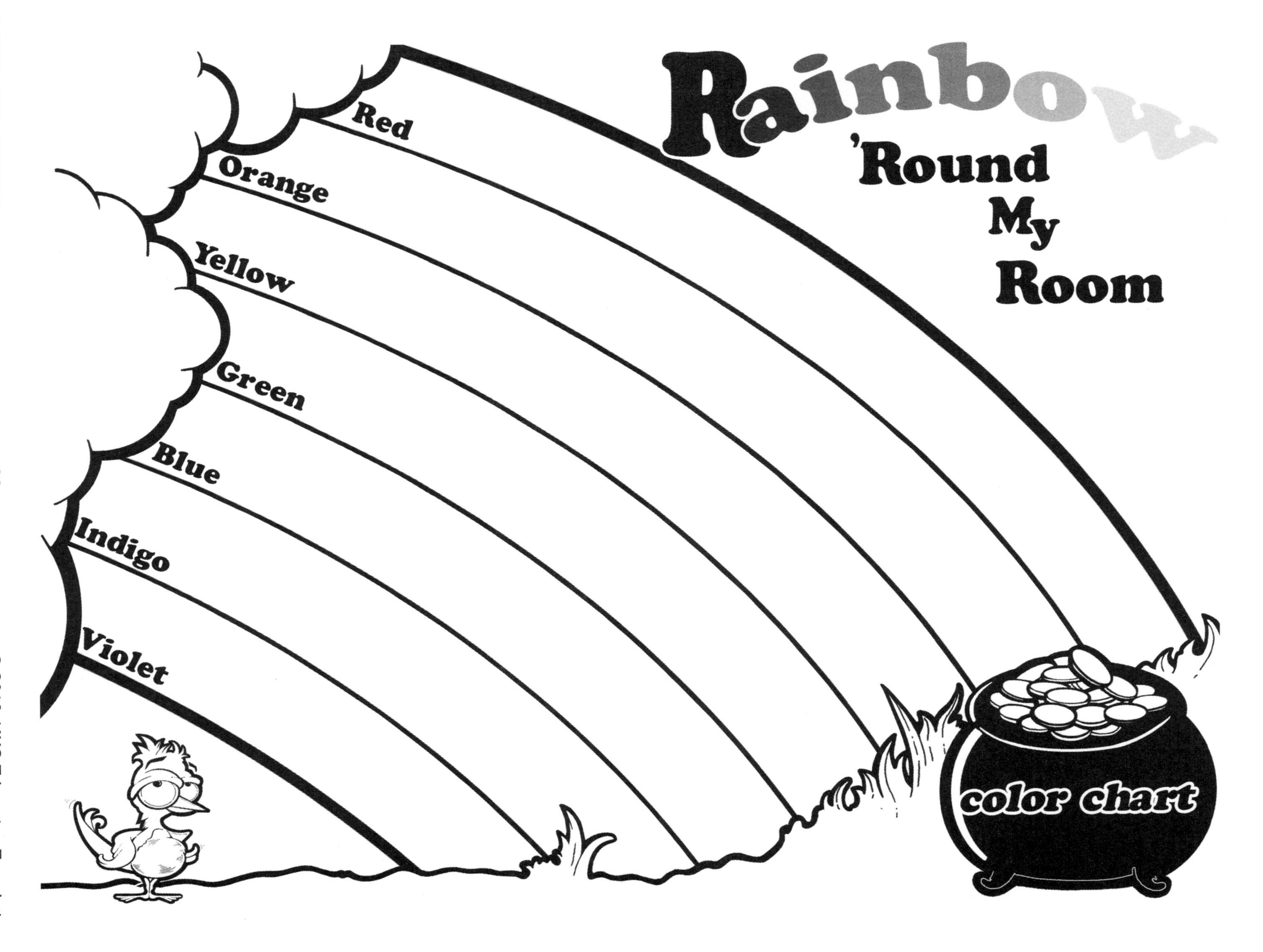
Rainbow
'Round
My
Room
Red
Orange
Yellow
Green
Blue
Indigo
Violet
color chart

Topic
Sense of sight

Key Question
How can we use the colors of blue, red, and yellow to make other colors?

Learning Goal
Students will become aware that they use their sense of sight to detect colors.

Guiding Document
Project 2061 Benchmarks

- *People can often learn about things around them by just observing those things carefully, but sometimes they can learn more by doing something to the things and noting what happens.*
- *People use their senses to find out about their surroundings and themselves. Different senses give different information. Sometimes a person can get different information about the same thing by moving closer to it or further away from it*

Science
Life science
human body
senses

Integrated Processes
Observing
Comparing and contrasting
Sorting and classifying
Collecting and recording data
Interpreting data

Materials
For the class:
crayons
6 cups prepared frosting (see *Management 3)*
food coloring (see *Management 3)*

For each student:
craft stick
paper plate
Color Combination sheet

Background Information
Teaching the recognition of color is one of the first concepts taught in primary classrooms. Tying color lessons to eyesight seems natural for students. In this activity students will mix red, blue, and yellow to produce orange, green, and purple. Their artistic expressions in color should begin to progress as they learn to mix colors to form other colors.

You may find students often perceive colors differently. Some of the students especially boys, may be colorblind or color deficient. Be sensitive to these differences and explain to students that we are all unique individuals. Going into the scientific information regarding the reasons for these differences would not be developmentally appropriate at this age. Simply learning that there are differences, or experiencing the differences, is enough.

Management
1. The activity has students mixing white cake frosting which has been colored red, blue, and yellow to discover the wide range of colors that can be made through combinations.
2. If using sugar in the classroom is a problem, softened cream cheese can be substituted for the cake frosting.
3. Prior to doing this activity, prepare frosting (or cream cheese). You will need a total of six cups of white frosting divided into three, two-cup containers. Paste food coloring that can be purchased at cake decorating stores or hobby supply stores will produce the best colors. Color two cups of the frosting red, two cups blue, and two cups yellow. You only need about ¼ teaspoon of paste food coloring for each container of frosting. Keep containers covered.
4. Prepare one paper plate per student by placing one tablespoon of each of the three colors of frosting (or cream cheese) around the rim of the plate leaving a lot of space between each color.

Procedure
1. Tell the students that they are going to learn how certain colors can be made.
2. Distribute a prepared paper plate and one craft stick to each student.
3. Ask the students to identify the colors of frosting on the plates. Tell them that these three colors will be used to make other colors.
4. Instruct the students to use their craft sticks to carefully separate a small portion of the red frosting and to put it in the middle of the plate. Have them add the same amount of yellow frosting to

the red sample and mix the two together. Ask them to identify the color made by mixing red and yellow. [orange]

5. Tell the students to lick their craft sticks so they do not contaminate the next colors they will mix.
6. Continue this method by mixing blue and yellow to make green and blue and red to make purple. Review the three original colors. Ask students to identify the colors made when two of these colors are mixed.
7. To reinforce the resulting colors, distribute the *Color Combinations* activity sheet. Have students color the top palettes with red and yellow, the middle palettes blue and yellow, and the bottom palettes blue and red. The intersections should produce orange, green, and purple respectively.
8. Have students compare and contrast their color combinations and ask if the colors they mixed are all the same. Students will notice that the colors are not the same. Tell them that the color depends on how much of each of the colors was mixed. Let the students have some free-exploration time to see what happens when more of one of the colors is added to the combination.
9. Distribute a second paper plate to each student. Inform them that they are now going to combine samples of their newly formed colors. Begin by combining equal amounts of orange frosting and green frosting. Continue combining green and purple and purple and orange from their first plate.

Connecting Learning

1. What color did you make when you mixed red and yellow? [orange]
2. Was your neighbor's orange the same? Why or why not? [it had more yellow, or it had more red]
3. How do the paint palettes you colored show the same thing?
4. What colors were made by mixing yellow and blue? [green] What are two ways you could show this to me? [mix the frosting or show you the colored palettes]
5. What happened when we mixed blue and red? [we get purple]
6. What were our original colors? [red, blue, yellow]
7. When we mixed two of these original colors, we formed another color. What colors did we form when we combined them? [orange, green, purple] (Remind students that they can check the *Color Combination* activity sheets they colored.)
8. What color did we get when we mixed orange and green?
9. Did we all get the same color? Why or why not? [different people used different amounts of colors]
10. Did anyone make black? If so, how did you do it?
11. What are you wondering now?

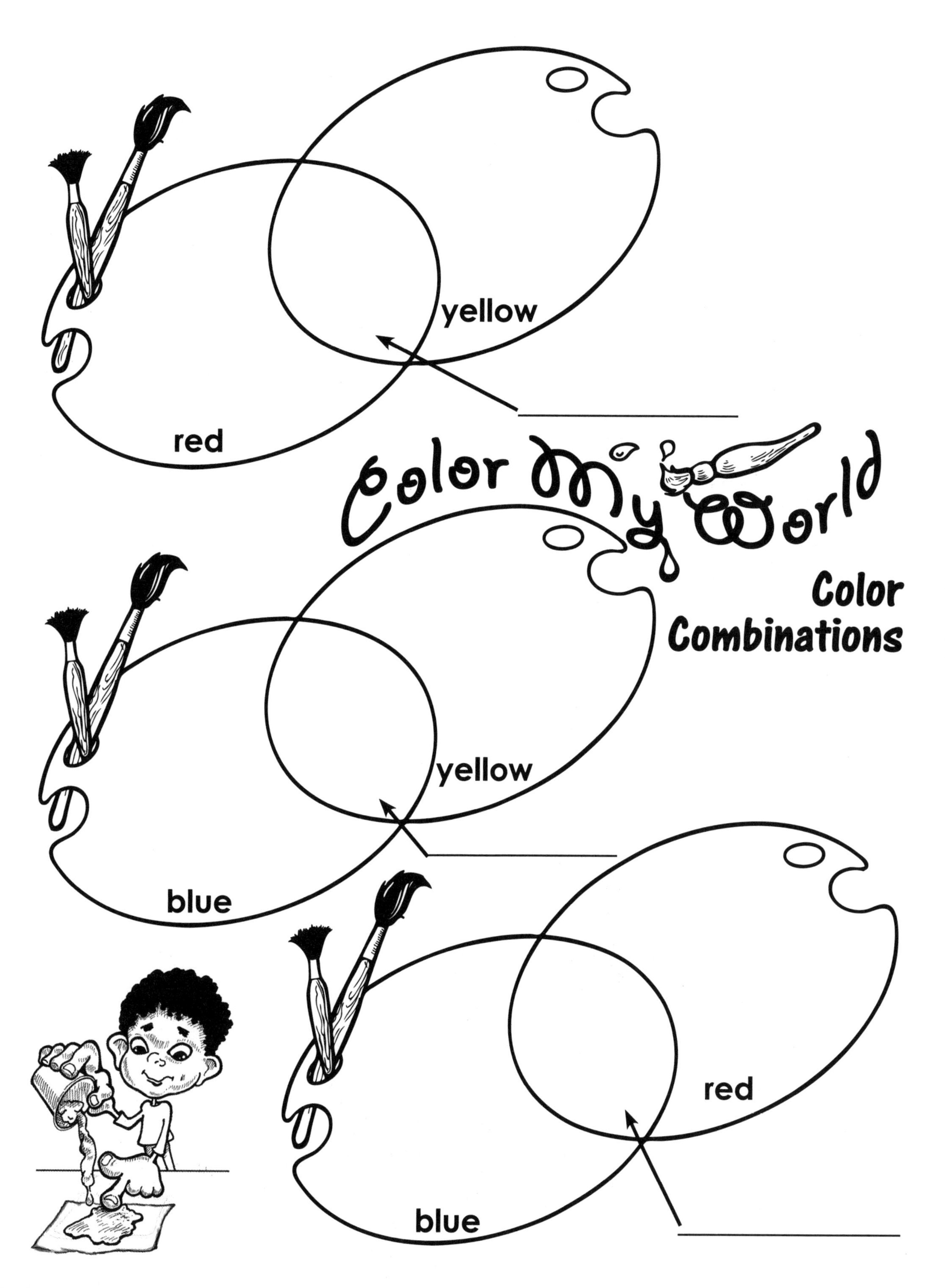

yellow
red
Color My World
Color Combinations
yellow
blue
red
blue

Topic
Sense of sight

Key Question
What does a color wheel show us?

Learning Goal
Students will become aware that they use their sense of sight to detect colors.

Guiding Document
Project 2061 Benchmarks
- *People can often learn about things around them by just observing those things carefully, but sometimes they can learn more by doing something to the things and noting what happens.*
- *People use their senses to find out about their surroundings and themselves. Different senses give different information. Sometimes a person can get different information about the same thing by moving closer to it or further away from it*

Science
Life science
    human body
        senses

Integrated Processes
Observing
Comparing and contrasting
Sorting and classifying
Collecting and recording data
Interpreting data
Communicating

Materials
For the class:
    Water Color Wheel (see *Management 1)*
    water
    food coloring: red, yellow, blue

For each group:
    3 paper or plastic cups, small
    3 drinking straws
    paper towels

For each student:
    wax paper
    2 *Water Color Wheel* activity pages
    toothpick
    crayons, optional

Background Information
Teaching the recognition of color is one of the first concepts taught in primary classrooms. Tying color lessons to eyesight seems natural for students. Bringing the color wheel into the lesson further enhances the lesson by reinforcing the connection between primary and secondary colors. The students' artistic expressions in color should begin to progress as they learn to mix colors to form other colors.

You may find students often perceive colors differently. Some of the students, especially the boys, may be colorblind or color deficient. Be sensitive to these differences and explain to students that we are all unique individuals. Going into the scientific information on the reasons for these differences would not be developmentally appropriate at this age. Simply learning that there are differences, or experiencing the differences, is enough.

Management
1. Enlarge the *Water Color Wheel* by using either an enlarging machine or an opaque projector. Use the appropriate colors to fill in the wedges of the *Water Color Wheel.*
2. Duplicate two copies of the *Water Color Wheel* activity page for each student.
3. This activity should be used as a follow-up to *Color My World.*
4. Put about 50 mL of water into each cup. Add food coloring so that each group will have a cup with red water, a cup with yellow water, and a cup with blue water.
5. Cut straws in half and put two halves in each cup.
6. Tear off wax paper so each student will have about one square foot.

Procedure
1. Introduce the *Water Color Wheel.* Ask students if they remember how the colors are made. [You begin with red, blue, and yellow. When you mix two of these, you get purple, orange, and green.]

2. Tell students that they are going to mix colors once again. Make certain each group of students has three cups of colored water with straws, wax paper, *Water Color Wheel* activity sheets, paper towels, and toothpicks.
3. Direct students to place the wax paper on top of one of their *Water Color Wheel* activity sheets. Show students how to put the straw into the colored water, squeeze the end of the straw tightly to hold the water in the straw, put the straw over the wax paper and to release the tight squeeze. Allow time for practice.
4. When students have learned the technique for putting drops of water onto their wax paper, have them use their paper towels to wipe all the drops off.
5. Have students put a drop the size of a nickel of each of the three primary colors onto their wax paper. Direct the students to place the appropriate color of water in the specified sections. Ask them to use their toothpicks to pull a little bit of the red drop to the center of the wax paper. Then have them pull a little of the yellow drop to the center and combine the two. Once orange is formed, have them drag that drop to the section between the red and yellow section labeled *orange* on their color wheel.
6. Continue as above with yellow and blue forming green and with red and blue forming purple.
7. Using their second *Water Color Wheel* activity sheet, direct the students to record the colors formed by putting drops of the colored water onto the sheet or using crayons. The dried, colored drops may not be as bright as the colored water, but students can easily distinguish the various colors. Have the students compare their results with the enlarged *Color Wheel* you made.
8. Using the circles at the bottom of the *Water Color Wheel* activity sheet, direct the students to drag opposite colors (complimentary colors) on the color wheel into the circles at the bottom of the page. For example: red and green in the first two circles. Then, as before, direct the students to drag a small portion of the red to the bottom circle, and then a small portion of the green into that same bottom circle. When these colors combine, they will form brown. Note: When you mix opposite colors on the color wheel together, you get different shades of browns. The more of one color you mix, the more the shade of brown has of that color. For example, if you mix equal amounts of red and green together, you will get brown; but if you mix more red into the sample, it will create a reddish-brown shade. More green in the mix produces a greenish brown.
9. Instruct the students to record their results, as before, on their second *Water Color Wheel* activity sheet.

Connecting Learning

1. How are our colors different? Are all our oranges the same color of orange? Are all our purples the same color of purple? Are the browns the same or different?
2. How do you think you make the different shades?
3. What colors did you mix together to get brown?
4. What happens when you mix red and yellow together? [You get orange, which is between red and yellow on the color wheel and in the rainbow.] What happens when you mix yellow and blue together? [You get green, which is between yellow and blue on the color wheel and in the rainbow.]
5. How do you make indigo and violet(purple)? [Indigo and violet (purple) are produced from the mixing of red and blue. Indigo has more blue than does violet (purple). (You may want to have students combine differing amounts of blue and red to see the various colors that are produced.)]
6. What are you wondering now?

Extension

Give the students old magazines, scraps of construction paper, and/or paint chip samples, which can be obtained from a local paint store. Ask students to find colors to put on the color wheel. You may want to discuss different shades and tints with the students.

Water Colors

Water Color Wheel

Violet

Red

Orange

Yellow

Green

Blue

Touch

Imagine brushing a feather across the tip of your nose or picking up a hot dish or getting slapped by a twig while walking in the woods. All these imagined events conger up memories that are associated with the sense of touch.

The sense of touch does not come from one specific location on the body. Your skin is studded with thousands of receptors that detect temperature, pressure, and pain. When contact is made with one of the receptors, messages called nerve impulses are sent along nerves to the brain. The brain receives the impulses and commands your body to respond. Over a period of time, some sense receptors may adapt to a certain feel so you no longer notice it.

Some areas of your body perceive touch more than others. The feather may tickle the end of your nose or your lips, but may not tickle your shoulders. The density of the receptors in your skin varies tremendously over the surface of your body. A square centimeter area on your fingertip may have dozens of skin receptors, while the same sized area on your back may have fewer than one.

Your skin provides a boundary between you and the external world. It is the largest sensory organ of your body. If you are a typical adult, your skin weighs about nine pounds and has a surface area of about 20 square feet.

Down Where the Waves Begin

We came to stay down at the beach,
My family and I,
To walk along its sandy shores
Beneath its brilliant sky.

The sand is gritty, damp and cold
Down where the waves begin;
I run to meet them as they come
And run away again.

And when I'm not quite fast enough,
They slap and sting and fizz;
I feel the wet sand slip away.
How interesting that is.

From all this sun and wind and play
My cheeks are hot and red.
"Enough of the outdoors for you,"
Is what my father said.

He scooped me up, and bundled me
Inside a rough, dry towel,
Then gathered all my playthings:
My pail, seashells and trowel.

Into the tub with you, my dear,
Clean clothes are on the shelf.
"There's nothing but pajamas here,"
I mutter to myself.

Then once I'm in the bathtub
With sand and ocean washed away,
As bubbles pop and tickle me,
It's where I want to stay.

But the last thing I remember
As I slip into my bed,
Is just how soft my pillow feels
Beneath my heavy head.

Brenda Dahl

written by Suzy Gazlay

How much can I touch,
Can I touch with my fingers?
I can touch my knees and ankles,
I can touch my toes;
I can touch my elbows,
I can touch my head and shoulders;
I can touch my middle,
I can touch my nose.

I touch furry, I touch fuzzy,
I touch soft or rough or pointed,
I touch slimy, I touch gooey,
I touch sand and dirt.
I touch cold, I touch warm,
Not too hot—my touch will warn me—
It's a way that touch protects me
So I don't get hurt.

It's funny how my brain
Is connected to my fingers,
'Cause no matter what I'm touching,
My brain can tell.
And I know they are connected,
'Cause my skin and brain together
Let me feel the world around me,
And they really do it well! . . .

- Words and parts of words that are underlined indicate need for a heavier accent in your voice.

Topic
Sense of touch

Key Question
How can you sort shapes using your sense of touch?

Learning Goals
Students will:
- sort and match shapes using their sense of sight,
- sort and match shapes using their sense of touch, and
- compare the two experiences.

Guiding Documents
Project 2061 Benchmarks
- *Numbers and shapes can be used to tell about things.*
- *People use their senses to find out about their surroundings and themselves. Different senses give different information. Sometimes a person can get different information about the same thing by moving closer to it or further away from it.*

*NCTM Standards 2000**
- *Sort, classify, and order objects by size, number, and other properties*
- *Sort and classify objects according to their attributes and organize data about the objects*
- *Recognize, name, build, draw, compare, and sort two- and three-dimensional shapes*
- *Recognize and represent shapes from different perspectives*

Math
Geometry
 2-D shapes
Sorting

Science
Life science
 human body
  senses

Integrated Processes
Observing
Comparing and contrasting
Recording
Communicating

Materials
For each group:
 sorting mat (see *Management 1)*
 shape card set (see *Management 2)*

For each student:
 blindfold (see *Management 3)*
 Feel and Find page

Background Information
We use information from previous experiences to help us understand our world and to learn about new things. Much of the information is gathered through the use of our five senses. In this activity, students will try to determine four different sandpaper shapes by using only their sense of touch. They will use their prior knowledge of shapes and the sensory input from touch to determine the shapes.

Management
1. Make a sorting mat for each group by dividing a 12" x 18" piece of construction paper into four sections as shown. You may want to put masking tape over the lines dividing the sections so students can determine the borders of each section on the mat using their sense of touch.

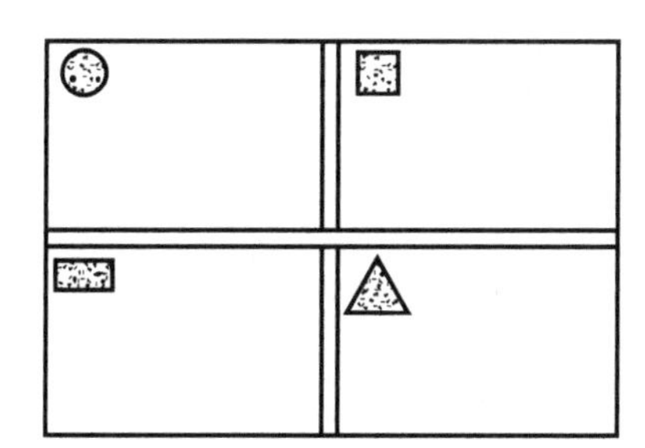

2. Cut the shapes included in this lesson from sandpaper. The shapes marked with an asterisk (*) will be glued onto the sorting mat as illustrated. Glue the remaining shapes to individual 3" x 5" cards that have been cut in half along the 5" dimension. Each group will need a sorting mat and a set of sandpaper cards.

3. To prevent the spread of any eye conditions from one child to another, it is advised that students use their own personal blindfolds. A pattern is included.
4. A cross-age tutor or adult aide is helpful in directing the search for shapes.

Procedure

1. As a whole class activity, show students the cards with the shapes on them. Ask them what they see. Ask them how they knew what the shapes were. What sense did they use in determining the shapes? [sense of sight]
2. Ask students how they might be able to identify the shapes if they could not use their sense of sight. [use the sense of touch]
3. Give each group a sorting mat and set of shape cards. Ask them to sort their cards using the shapes on the mat as a guideline. Discuss whether or not this was difficult and why.
4. Tell the students they will each have a turn to go to a station and, with a blindfold over their eyes, they will classify the shape cards by placing them in the appropriate sections of their sorting mats.
5. Have each group bring its set of shape cards and sorting mat to the station so that there are enough sets for all the members of one group to participate at the same time.
6. Ask groups of students to bring their personal blindfolds to the station when it is their turn. Allow one group to work at the station at a time.
7. While blindfolded, let each student feel the shape on each card and put it into the appropriate section of his or her sorting mat.
8. After students have finished, have them remove the blindfolds and draw their results on the *Feel and Find*. Instruct them to draw the shapes they classified in each section, whether right or wrong.

Connecting Learning

1. What sense did you use to sort the shapes the first time? [sight] Was this easy or difficult? Explain.
2. What sense did you use to sort the shapes the second time? [touch] Was this easier or harder than using the sense of sight? Explain.
3. Did you make any errors in classifying the shapes when you were using your sense of touch?
4. Which shape was the easiest for you to classify? Why? Which shape was the most difficult? Why?
5. Can you think of times where you rely only on your sense of touch to find or identify something? [locating something in the dark, searching for an object inside a bag or backpack without looking, etc.]
6. What are you wondering now?

Extensions

1. While the students are blindfolded, direct them to make a pattern with the cards. [circle, square, circle, square... circle, square, rectangle, circle, square, rectangle...]
2. While the students are blindfolded, let them feel objects such as blocks, balls, or erasers, and determine what they are.
3. Add a third-sized shape to introduce the terms big, bigger, biggest or small, smaller, smallest. Repeat the activity, sorting by size instead of shape. Find the biggest circle, the smallest square. Which of these two shapes is bigger? How can you tell?

Home Links

1. Give students a homework paper like the sorting mat. Have them draw something in each section from their home that they could tell the shape of by touching it. A square table would be acceptable, but squares on non-textured wallpaper would not be acceptable because students would need their eyes to determine the shape.
2. Have each student share an object that is one of the four shapes studied in this lesson. Set it at the station for students to touch.
3. Have students bring a picture of an object with one of the shapes and paste it on a large class poster that has been divided into sections like the sorting mat.

* Reprinted with permission from *Principles and Standards for School Mathematics,* 2000 by the National Council of Teachers of Mathematics. All rights reserved.

Patterns for Student Blindfolds

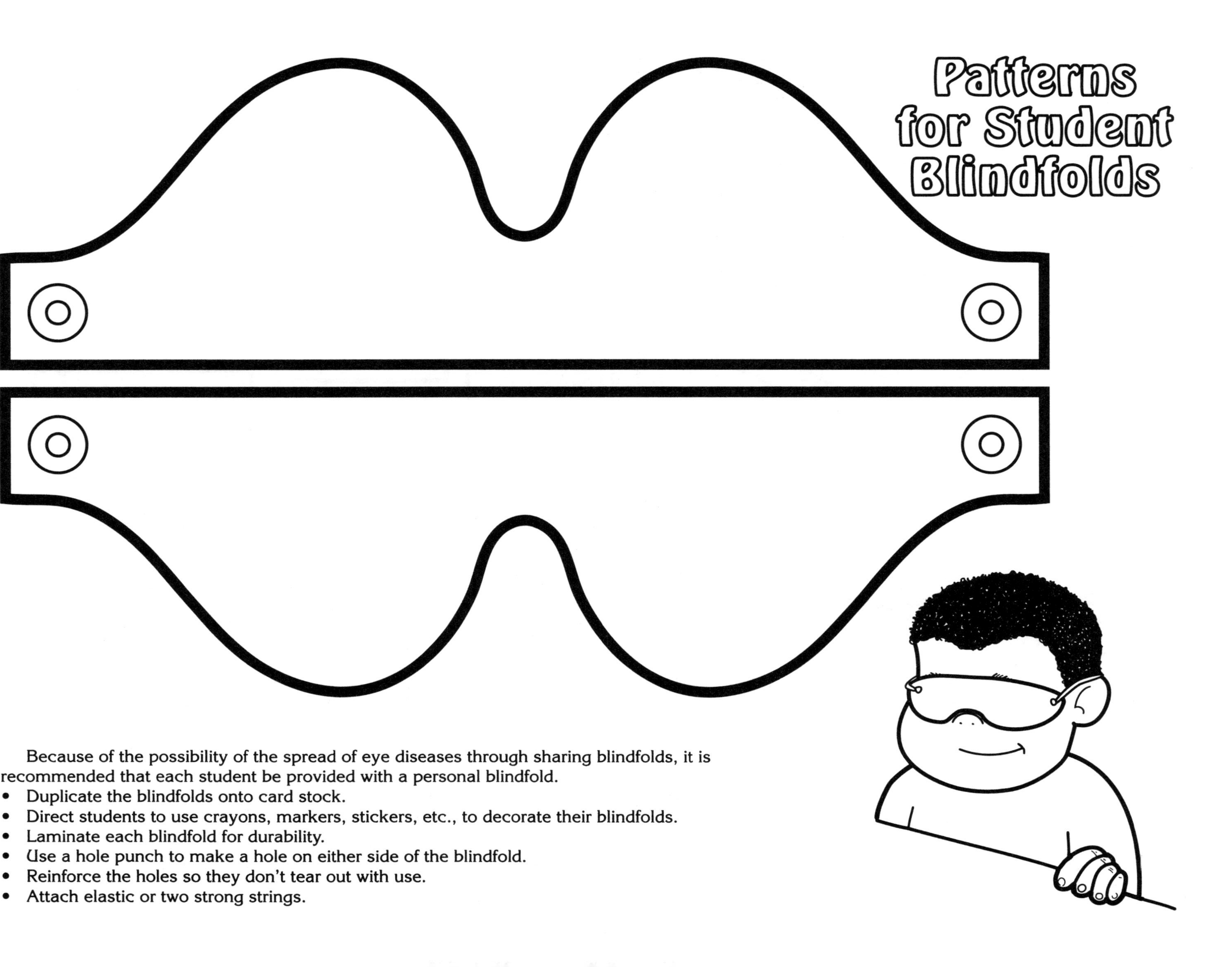

Because of the possibility of the spread of eye diseases through sharing blindfolds, it is recommended that each student be provided with a personal blindfold.

- Duplicate the blindfolds onto card stock.
- Direct students to use crayons, markers, stickers, etc., to decorate their blindfolds.
- Laminate each blindfold for durability.
- Use a hole punch to make a hole on either side of the blindfold.
- Reinforce the holes so they don't tear out with use.
- Attach elastic or two strong strings.

Cut these shapes out of sandpaper for each group.

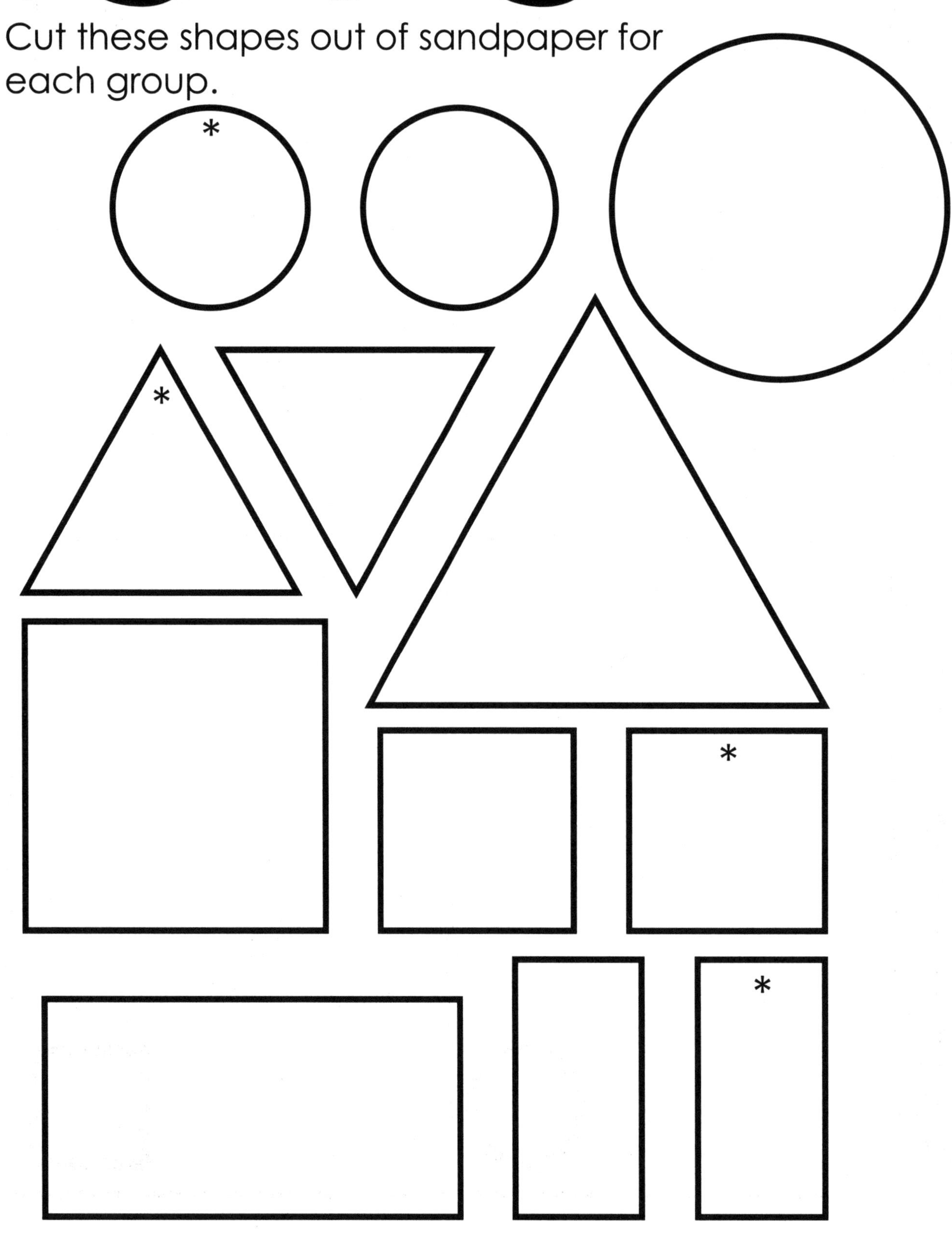

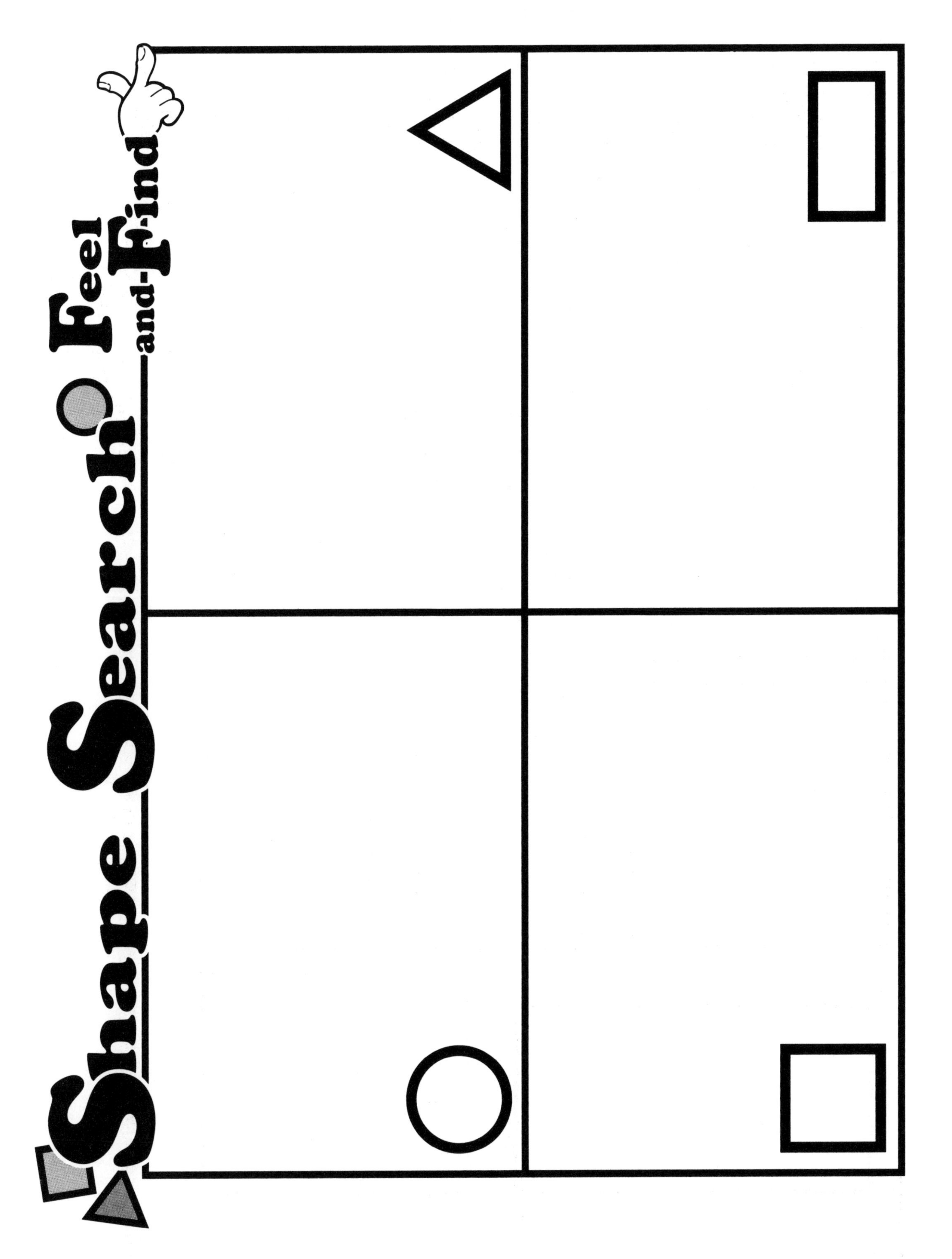
Shape Search
Feel-and-Find

Topic
Sense of touch

Key Question
How can you match the textures of pieces of fabric without seeing them?

Learning Goal
Students will identify and classify rough and smooth textures by using the sense of touch.

Guiding Documents
Project 2061 Benchmark
- *People use their senses to find out about their surroundings and themselves. Different senses give different information. Sometimes a person can get different information about the same thing by moving closer to it or further away from it.*

*NCTM Standards 2000**
- *Sort and classify objects according to their attributes and organize data about the objects*
- *Represent data using concrete objects, pictures, and graphs*

Math
Graphing
Tallying

Science
Life science
human body
senses

Integrated Processes
Observing
Comparing and contrasting
Sorting and classifying
Predicting
Recording data
Interpreting data
Communicating

Materials
10 plastic cups, 9 oz
5 pairs adult-size tube socks
Fabric swatches (see *Management 2)*
Card stock, 3-inch squares (see *Management 3)*
Glue
Scissors
Tally sheet, one per student
Sorting sheet (see *Management 4)*

Background Information
The sense of touch is not highly developed in young children. In an attempt to further develop this sense, the teacher is encouraged to explore many experiences that use the sense of touch. In order for children to verbalize about what they are touching, it is first necessary to define the terms of classification. Because these terms are relative, it is often difficult for students to apply them to various objects. This activity has students classify fabric swatches as rough or smooth. A beginning definition for rough may be having a bumpy-type surface like the feel of carpet or sandpaper. Smooth things have a slick-type surface like a mirror or the glass on the window. The teacher may wish to do a series of introductory lessons classifying items in the classroom into rough and smooth categories. Once the categories have been defined and experienced, this lesson can be utilized.

Management
1. Prepare the *Touch and Tell* cups by inserting the cups into the toes of the tube socks, one cup per sock.
2. Select at least 10 kinds of fabrics that can be easily distinguished as rough or smooth. Suggested types include velvet, burlap, dotted Swiss, linen, cotton, satin, silk, and nylon netting. Cut the fabric into 3" x 3" squares with two squares of each type.
3. To prepare the touch cards, glue the fabric swatches to the squares of card stock. Place one card of each fabric type inside a *Touch and Tell* cup and leave the second card of each fabric type out.
4. Duplicate one sorting sheet for use at the *Touch and Tell* center for the students to sort and classify their "textured pairs" (the matching fabric swatch cards). For durability, copy the sorting sheet on card stock and laminate.
5. The tally sheets come four to a page. Make enough copies so that each student can have one.
6. This activity can be done as a learning station activity, with groups of eight to 10 students exploring the various textures.

Procedure

1. Show the *Touch and Tell* cups to the children. Explain that they will be using their sense of touch to find textures that are the same and they will then place the "textured pairs" in either the *smooth* or *rough* categories on the sorting sheet.
2. Hold up the sorting sheet and have the students review what smooth and rough mean by locating several objects in the room that would represent smooth and rough.
3. Ask how they can use the sense of touch to find textures that are the same.
4. Encourage the students to touch some of the fabric squares that are not inside the socks.
5. Choose one student to reach inside a *Touch and Tell* cup and, through the sense of touch, determine which fabric square would be a match to the one in the cup. Direct the student to point to the square that he/she predicts is a match before pulling the square out of the cup.
6. Direct all the students to take turns using the *Touch and Tell* cups until all pairs are matched.
7. As each fabric pair is found, have the students place the fabric pair in the appropriate category on the sheet.
8. Using the student tally sheets, tally to find out which category has the most/least.

Connecting Learning

1. What sense did we use today to find the matching pairs?
2. If you had a difficult time sorting the squares, what other sense would make the task easier?
3. What group, smooth or rough, had the most fabric pieces?
4. When you think of smooth things, what do you think of?
5. What things can you name that are rough?
6. Look closely at the rough fabrics. How are they alike? How are they different from any of the smooth fabrics? What does this have to do with making them smooth or rough?
7. Explain why you can use just your sense of touch to sort the squares into rough or smooth.
8. Explain why you cannot use just your sense of touch to sort the squares by color.
9. Using your tally sheets, tell me how many things that you touched were rough. ...smooth.
10. What are you wondering now?

Extensions

1. Add other swatches and repeat the activity.
2. Use pairs of similar shaped objects and challenge the students to match pairs by touch.
3. Have students classify the objects in their backpacks into smooth and rough categories.
4. Change the categories from smooth and rough to hard and soft.
5. Rearrange the fabric squares into different attributes, e.g., color, pattern design, etc. Would the rules about matching our squares need to change? Could we sort with only our sense of touch?
6. Build a classroom rock collection for purposes of sorting and classifying. Begin with smooth and rough and then substitute attributes such as shiny, dull, speckled, striped, etc.

Curriculum Correlation

Language Arts

Make a *Things That Are Rough and Things That Are Smooth* book for your classroom library. Have each student draw an illustration for each category and share it with classmates before putting it into the book.

Home Links

1. Ask students to collect and record three objects that would fit into each category, smooth and rough.
2. Encourage the parent/child to send real objects for a "Texture Sharing Day." Emphasize only the two textures discussed, and tally
   a. the total number of objects brought from home,
   b. the total number of smooth-textured objects, and
   c. the total number of rough-textured objects.

Touch and Tell
Sorting Sheet

silk

sandpaper

smooth	rough

Touch and Tell
Tally Sheets

Topic
Sense of touch

Key Question
What is the texture of these objects?

Learning Goal
Students will use their sense of touch to sort and classify objects by texture.

Guiding Documents
Project 2061 Benchmarks
- *People use their senses to find out about their surroundings and themselves. Different senses give different information. Sometimes a person can get different information about the same thing by moving closer to it or further away from it*
- *Objects can be described in terms of the materials they are made of (clay, cloth, paper, etc.) and their physical properties (color, size, shape, weight, texture, flexibility, etc.).*

NRC Standards
- *Ask a question about objects, organisms, and events in the environment.*
- *Objects are made of one or more materials, such as paper, wood, and metal. Objects can be described by the properties of the materials from which they are made, and those properties can be used to separate or sort a group of objects or materials.*

*NCTM Standard 2000**
- *Sort and classify objects according to their attributes and organize data about the objects*

Science
Life science
  human body
    senses

Integrated Processes
Observing
Comparing and contrasting
Sorting and classifying
Predicting
Collecting and recording data
Interpreting data

Materials
Chart paper or bulletin board paper
Newsprint (see *Management 3)*
Old crayons
Various objects with distinctive textures
  (see *Management 6)*
Scissors
Glue
3" sticky notes
Student page copied on card stock

Background Information
Touch gives people information about objects in the world around them. Size, shape, and texture are properties that can be sensed through touch. Nerve endings in the skin called sense receptors allow people to receive the sensations. These sense receptors are found in all three layers of the skin. The receptors send messages about what is felt through nerve impulses, by way of the nervous system, to the brain.

Key Vocabulary
Texture: The characteristic structure of a surface
Rough: A surface that is uneven as a result of projections, irregularities, or breaks
Smooth: A surface that is relatively free from projections and irregularities

Management
1. Students can work on sorting different objects individually or within small groups after some discussion about *texture* and the terms *rough* and *smooth.*
2. Keep the students focused on the qualities of roughness and smoothness during this investigation. Later, other classifications may be added.
3. The procedures for rubbings should first be modeled and then supervised briefly to ensure understanding of the process. Newsprint needs to be placed over the object to be rubbed. Remove the paper from an old crayon and, using its broad side, rub it sideways several times over the object until the imprint of the object appears. While rubbing with one hand, it may be necessary to keep the other hand on top of the paper and object so they do not move. Students may need to practice this several times with different objects before getting a satisfactory rubbing. The students may want to create a picture or collage of the rubbings.

4. Make a class chart or a class mural for recording objects found in the room by the students.
5. Copy the student page onto card stock for each student.
6. You will need a collection of several objects with distinctive textures. Suggested items are: corrugated cardboard, Velcro™, plastic wrap, wax paper, velvet, leaf, corduroy, sandpaper, etc. Students will first sort and classify these items. Later, they will glue samples of these items to their student pages.

Procedure

1. Set out objects with rough or smooth textures. Let students verbalize how the objects feel. Guide them to concluding that all things have texture. Lead them into sorting these objects into categories of *rough* and *smooth*.
2. Encourage students to point out other objects in the room that they predict would fit into the categories of *rough* and *smooth*. (Do not allow students to feel these objects before predicting.)
3. Record students' predictions on chart paper.
4. Collect the classroom objects identified by the students and ask the students to feel the surfaces. Compare and contrast the actual results with those predicted.
5. Go outside and repeat this activity finding various objects with smooth or rough characteristics.
6. Select several objects the students have collected and let them arrange those objects from the roughest to the smoothest.
7. Distribute the student page to students. Instruct them to select three items that are rough and three items that are smooth from the collection of objects. Tell them to glue each of these items to the sections on the paper, using one item per section. Instruct them to use scissors, if necessary, to cut things down to size so that they fit within the white portions of the squares.
8. Once the glue has dried, give each student six sticky notes. Instruct them to place the sticky notes over the items. Be sure they put the sticky part of the note over the gray box at the top of each square.
9. Have students share their flap pages with others and identify each texture as rough or smooth.

Connecting Learning

1. How do the objects feel? What words describe their textures?
2. What sense did we use to feel the objects' textures?
3. How are the objects the same? How are they different?
4. Are some objects rougher or smoother than others? How can you tell?
5. Can you tell the texture of objects by looking at them? Explain.
6. How many rough objects did we find? How many smooth objects?
7. Did we find more rough or more smooth objects? How could you tell?
8. When would you want to use something that is rough?
9. When would you want something smooth?
10. What are you wondering now?

Extensions

1. Choose different texture characteristics to observe and record.
2. Use different parts of the body [feet, hair, skin] to decide their textures.

* Reprinted with permission from *Principles and Standards for School Mathematics*, 2000 by the National Council of Teachers of Mathematics. All rights reserved.

Attach sticky note here.

Glue object here.

Rough or Smooth?

Attach sticky note here.

Glue object here.

Rough or Smooth?

Attach sticky note here.

Glue object here.

Rough or Smooth?

Attach sticky note here.

Glue object here.

Rough or Smooth?

Attach sticky note here.

Glue object here.

Rough or Smooth?

Attach sticky note here.

Glue object here.

Rough or Smooth?

Topic
Senses of touch and sight

Key Question
What senses do you use when you pick up an object?

Learning Goal
Students will gain an understanding of how important the senses of touch and sight are to the understanding of our world.

Guiding Documents
Project 2061 Benchmark
- *People use their senses to find out about their surroundings and themselves. Different senses give different information. Sometimes a person can get different information about the same thing by moving closer to it or further away from it.*

*NCTM Standards 2000**
- *Count with understanding and recognize "how many" in sets of objects*
- *Sort and classify objects according to their attributes and organize data about the objects*
- *Represent data using concrete objects, pictures, and graphs*

Math
Graphing

Science
Life science
  human body
    senses

Integrated Processes
Observing
Comparing and contrasting
Predicting
Collecting and organizing data
Interpreting data

Materials
For the class:
  chart paper

For each group:
  one pair of women's heavy-duty work gloves
    or child-size knitted gloves
  10 cotton balls (see *Management 1*)

For each student:
  crayons
  personal blindfold
  student journal (see *Management 4)*
  Kid Gloves graph

Background Information
To realize the value of their sense of touch, students will diminish that sense by wearing gloves. To further mask the ability to feel, you must also take away sight with the use of a blindfold so that the students may not see what they are trying to pick up. This activity demonstrates to the students how their sense of touch helps them interact with their environment through picking up objects.

Management
1. Set up a station with a pair of gloves and 10 cotton balls. Use inexpensive cotton balls! They will be harder for students to feel.
2. If students do not have personal blindfolds, make these prior to doing the activity.
3. Students will work in a small group at this station while other groups are involved at other stations. An adult helper or cross-age tutor is needed for this activity.
4. Copy and construct student journals prior to classtime. To construct the journal, fold the page in half lengthwise and then widthwise so that the text is on the outside and the pages go in order.

Procedure
1. Ask students if they have ever tried to find anything in the dark. Have them describe their techniques for doing this. [feel the wall to find the light switch; walk with hands outstretched to find the door; shuffle feet to find slippers or shoes]
2. Show the students a cotton ball. Discuss its attributes. Ask them if it would be difficult to pick it up. Give students some free exploration time holding cotton balls, picking them up, and setting them down.

3. Discuss what senses they would use to pick up the cotton ball.
4. Tell students they are going to form a prediction about picking up 10 cotton balls. Write on chart paper and say, "If I can see and feel the cotton balls, I will be able to pick up ____." (Most students will say all 10.)
5. Ask two or three students to test the prediction for the class. Have them show the rest of the class that if they use their senses of sight and touch, they can pick up all 10 cotton balls.
6. Tell students that they are going to see what happens when they don't use their senses of sight and touch. Ask them what they could do so they don't use their sense of sight. [wear a blindfold, close our eyes]
7. Inform students that they will form a prediction about not being able to see the cotton balls. "If I cannot see the cotton balls, I will be able to pick up ____." Distribute the *Kid Gloves* graphs. Have students draw enough cotton balls to represent their prediction of the number (out of 10) they think they can pick up while wearing a blindfold. Students will test this prediction at the center.
8. Ask students what would happen if they couldn't see and couldn't feel the cotton balls. Have the students discuss what they could do so that they don't use their sense of touch, but still pick up the cotton balls. Lead them to conclude that if they wore heavy gloves, they couldn't feel the cotton balls.
9. Have them form a prediction about how many cotton balls (out of 10) they could pick up without seeing or feeling them and draw their predictions in the appropriate place on the graph.
10. Divide students into groups. One group will go to the *Kid Gloves* station while the others will go to other stations.
11. Have students at the station count the cotton balls to confirm that there are 10. Ask them to state their predictions. [If I cannot see the cotton balls, I will be able to pick up ____.] One at a time, have students put on their blindfolds to test their predictions.
12. After determining how many cotton balls were picked up, have each student draw in cotton balls to represent the actual results on the graph.
13. Remind students of the other prediction they formed about not being able to see or feel. [If I cannot see or feel the cotton balls, I can pick up ____.]
14. Have students test this prediction by wearing their blindfolds and gloves. After they have performed the test, direct them to record the actual results on their graphs. Also have them record their predictions and actual results in the student journal to take home.
15. Continue until all the groups have completed the activities at this station.

Connecting Learning
1. How close were your predictions to the actual results? Explain how you know.
2. Why was the test difficult? What would have made it easier?
3. Which test was the most difficult? Why do you think so?
4. Would any other senses affect the data? Explain.
5. Even though you can see them, are there times you need to feel things to find out more information about them? [Sometimes you cannot tell if something is warm or cold by looking at it.]
6. What are you wondering now?

Extensions
1. Repeat the activity with gloves but without using the blindfold. Does it make a difference?
2. Repeat the activity with a blindfold, but use different objects (feathers, blocks). How much difference does the shape and weight of the object make?
3. With blindfolds on, give students an object and ask them to identify it.
4. Without allowing students to touch an object, ask them to describe what it feels like (hard, soft, cold, warm, slimy, wet). Why is this hard to do?
5. Try this activity again with cotton balls sprayed with perfume. What additional sense are you now using to find the objects to pick up when you are blindfolded and have gloves on?
6. Use small round jingle bells instead of cotton balls. What additional sense are you now using to find the objects to pick up when you are blindfolded and have gloves on?

Home Links
1. Give students the graph page and four cotton balls to take home. Have them try the activity on family members and record the numbers picked up by each person tested.
2. Have students try again at home using objects of different weights. Are the results different?

Kid Gloves

How many cotton balls can I pick up?

My Own Journal

1

Prediction

2

Actual

3

How many cotton balls did I pick up?

4

Topic
Sense of touch

Key Question
Without using your sense of sight, how can you match a given bead pattern?

Learning Goal
Students will use their sense of touch to complete a given pattern made out of various shaped beads.

Guiding Documents
Project 2061 Benchmark
- *People use their senses to find out about their surroundings and themselves. Different senses give different information. Sometimes a person can get different information about the same thing by moving closer to it or further away from it.*

*NCTM Standards 2000**
- *Sort, classify, and order objects by size, number, and other properties*
- *Recognize, describe, and extend patterns such as sequences of sounds and shapes or simple numeric patterns and translate from one representation to another*

Math
Patterning
Tallying

Science
Life science
human body
senses

Integrated Processes
Observing
Comparing and contrasting
Predicting
Collecting and recording data
Interpreting data

Materials
Collection of large beads of different shapes
One paper bag per student at the station
String
One tally sheet for each student

Background Information
Exploring, duplicating, and creating patterns is an important strategy in problem solving. In science and mathematics, we try to find solutions to problems by studying patterns and searching for clues. Student exploration and creation of patterns enables them to better understand and analyze patterns.To begin this exploration and creation of patterns, we begin by looking at patterns made by using an obvious attribute such as shape.

Pattern, as it pertains to this activity, is defined as *a repeated arrangement of shapes in a linear order.* It is assumed that students will have had multiple experiences in patterning prior to this lesson in which they will be duplicating patterns of beads in a variety of shapes held together on a string. Using only their sense of touch to search for the needed shape, the students will duplicate a given pattern by reaching into a paper bag to find the specific shapes.They will need to search for each shape in the order in which it occurs in the sample pattern.

The students will gain further understanding of the sense of touch in real-world experiences. They will also gain insight into the use of touch by persons who have lost their eyesight.

Management

1. It is suggested that you conduct this lesson at a center or station.
2. Prior to the lesson, string various-shaped beads into repeated patterns. Use approximately eight beads per string. Make enough for one bead pattern per student at the station. Attribute beads with lacing holes (item number 4120) are available from AIMS.
3. Use bead patterns consisting of three of four different shapes: cylindrical, round, square, ovate. Arrange them in patterns such as round, round, square, round, round, square, round, round, square; square, ovate, cylindrical, cylindrical, square, ovate, cylindrical, cylindrical; etc.

4. For each bead pattern, prepare a paper bag containing a string and a duplicate set of the beads you used on the bead patterns.
5. Prepare a tally sheet for each student.

Procedure

1. Show the students the bead patterns you have previously made.
2. Tell them that they will need to make another bead pattern just like the ones in front of them.
3. Tell them that the materials to do this are in the paper sack in front of them, but that they cannot look inside. Ask the *Key Question*.
4. Once the students decide that they can use their sense of touch to distinguish the beads, allow them to begin.
5. Through trial and error, the students will experience how to correctly find the shapes they are looking for to duplicate the patterns.

6. Instruct the students to keep a tally of how many correct and how many incorrect attempts they made to find the correct shapes.

Connecting Learning

1. Why was it difficult to find the shapes you needed? Which shapes were the easiest? ...the hardest?
2. Could you use specific colors using only your sense of touch? What sense would you need to use to do this?
3. Would it be more difficult or less difficult to identify the beads using your toes rather than your fingers? Why?
4. What are you wondering now?

Extensions

1. Allow students to create their own patterns. Have students trade with a partner and try to duplicate each other's patterns, using the same touch technique in the paper bag.
2. Allow students to try to duplicate the patterns using their toes to find the needed shapes.

* Reprinted with permission from *Principles and Standards for School Mathematics*, 2000 by the National Council of Teachers of Mathematics. All rights reserved.

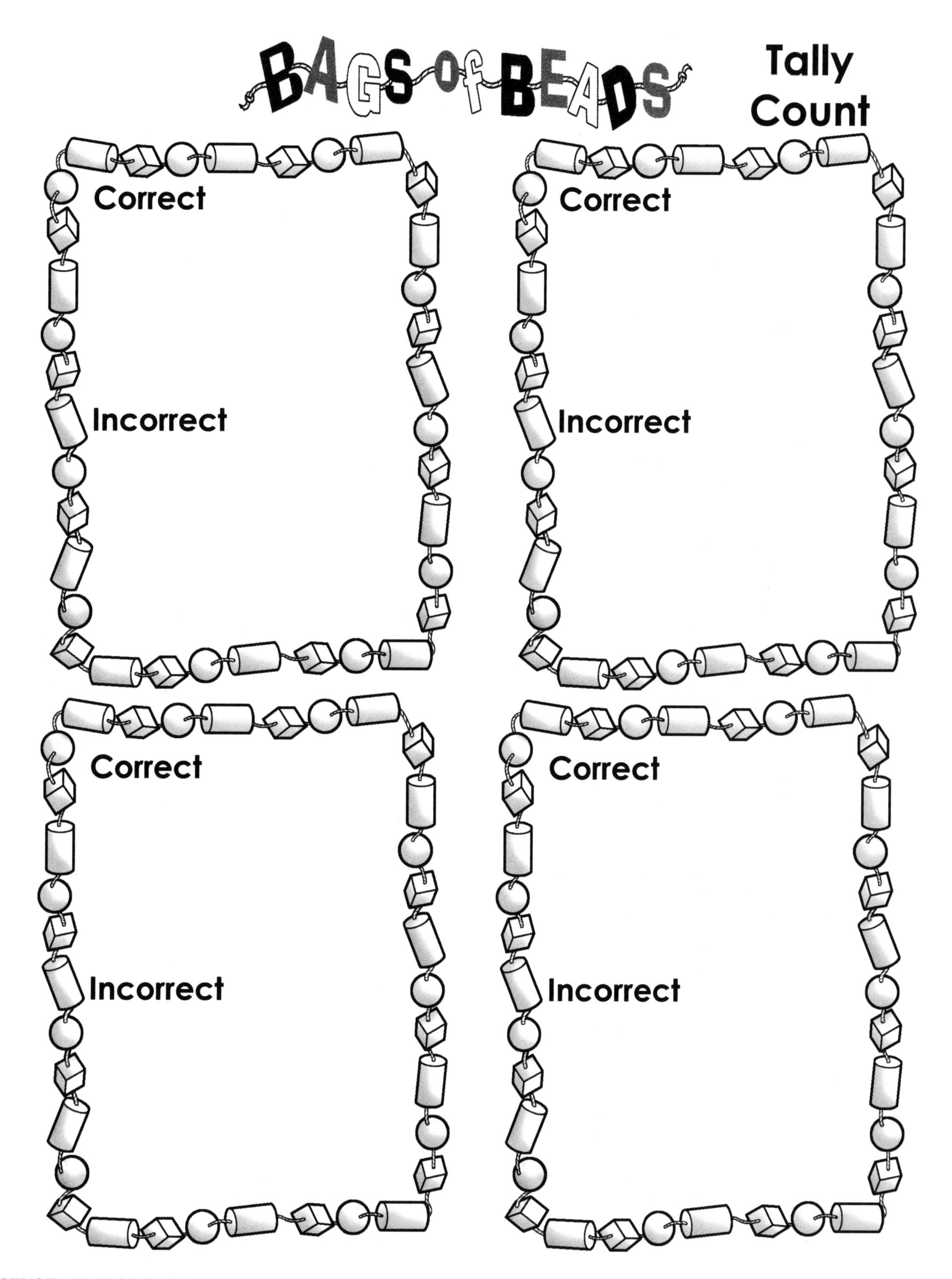
BAGS of BEADS
Tally Count
Correct
Incorrect
Correct
Incorrect
Correct
Incorrect
Correct
Incorrect

Topic
Sense of touch

Key Question
Are your feet or your hands more sensitive to the touch of a feather?

Learning Goal
Students will determine which parts of their body, their feet or their hands, are the most sensitive to the touch of a feather.

Guiding Documents
Project 2061 Benchmark
- *People use their senses to find out about their surroundings and themselves. Different senses give different information. Sometimes a person can get different information about the same thing by moving closer to it or further away from it*

*NCTM Standard 2000**
- *Represent data using concrete objects, pictures, and graphs*

Math
Graphing

Science
Life science
  human body
    senses

Integrated Processes
Observing
Comparing and contrasting
Predicting
Collecting and recording data
Interpreting data

Materials
Feathers
Class graphs
Graph markers

Background Information
The skin is the major receptor of tactile sensation. We all enjoy the fun and laughter that comes from a good tickle. The tickle is the result of a soft touch to a very sensitive part of our body. This activity uses a feather to determine whether the feet or the hands are more sensitive.

The sensitivity to touch may vary among humans because of various lifestyles. Those students who have gone barefooted more than others will be less sensitive to the tickle of a feather on their feet than those who rarely go barefooted. The skin on the bottom of the feet of those who go barefoot will callous more than the feet of those who always wear shoes. When the skin is calloused, the touch receptors are further from the surface and thus are not as sensitive to touch.

Management
1. Enlarge the class graph on the board or on a piece of chart paper.
2. Cut out two hand and two foot graph markers for each student.

Procedure
1. Ask the students to remove their shoes and socks. Tell them to find a partner with whom to work.
2. Ask the *Key Question*. Have students show their predictions by placing the appropriate graph markers on the class prediction graph.
3. Direct one student in each pair to gently tickle the palm of the other student's hand, then the sole of the same student's foot.
4. After lots of giggles and jiggles, ask the students who were tickled to record the actual results on the graph of which area was the more ticklish.
5. Have students switch roles and follow the same procedure of tickling and recording actual results.
6. After discussing the results, allow students to test other parts of their bodies: elbows, the tops of their hands, the tops of their feet, the ear lobes, their noses, etc.

Connecting Learning
1. Were all of us the most ticklish on the bottoms of our feet? How do you know?
2. Why were some of us more ticklish on the palms of our hands?
3. What sense were we using to feel the tickles?
4. What does our graph tell us about whether our hands or our feet are more sensitive to the touch of a feather?
5. What did you find out about other parts of your body?
6. What are you wondering now?

* Reprinted with permission from *Principles and Standards for School Mathematics*, 2000 by the National Council of Teachers of Mathematics. All rights reserved.

Which is more ticklish?

I think	I think	I learned	I learned

You Tickle My Fancy

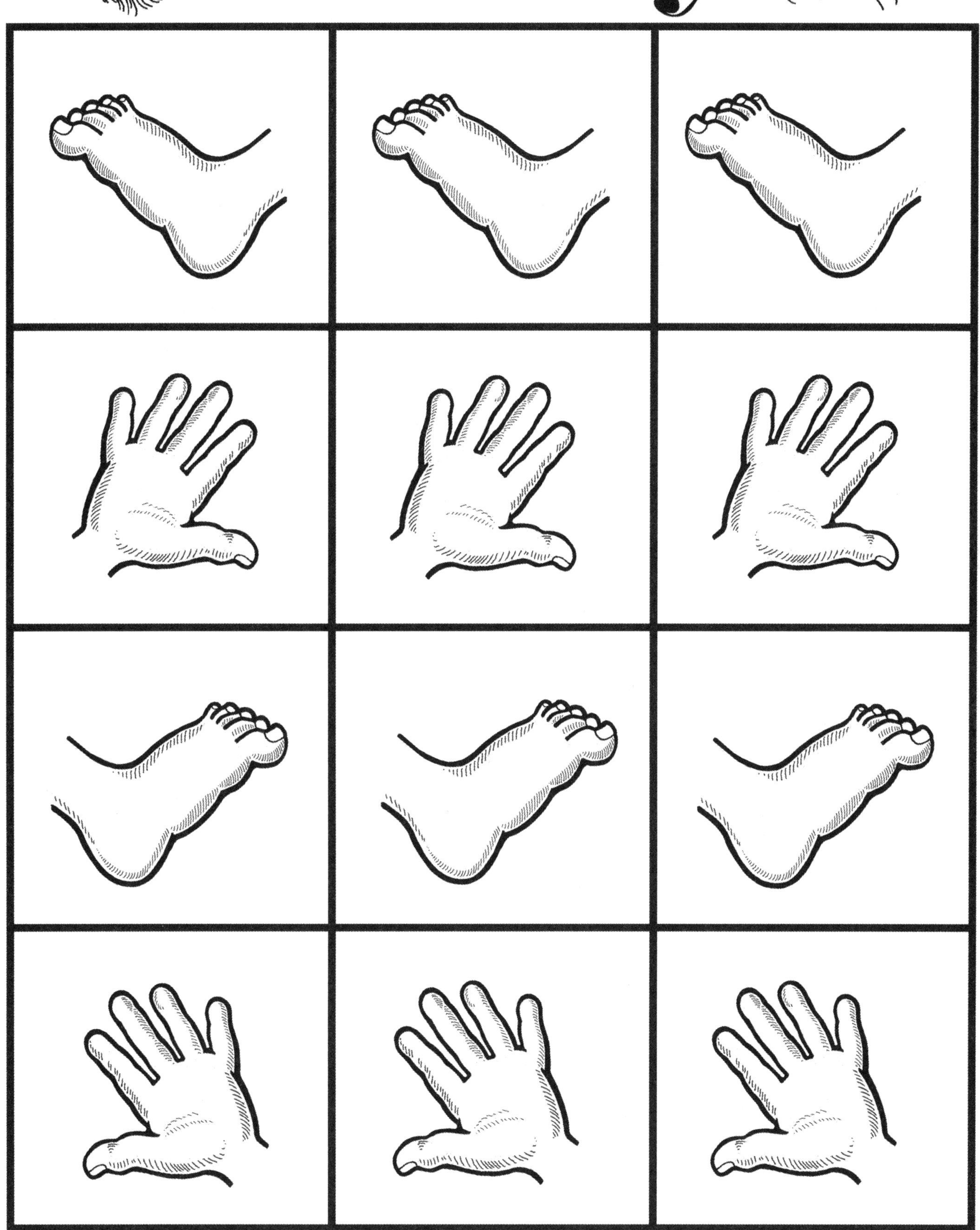

Taste

Our ability to taste begins in our mouth. Our tongues are covered with thousands of tiny receptors called taste buds that are housed in the bumps (papillae) that cover the tongue. These taste buds contain taste cells, which in turn have microscopic hairs called microvilli. These microvilli send messages to the brain by means of nerves. The brain interprets the messages, along with messages from the olfactory receptors in the nose, to identify what is being tasted.

Humans are born with around 10,000 taste buds, but as we get older, the number of taste buds decreases. One problem facing those who care for the elderly is getting them to eat. Their lack of appetite is often due to their inability to detect the flavor of food; therefore, they may lose interest in eating.

For many years, western scientists recognized only four basic tastes—sweet, sour, salty, and bitter. In recent years, however, a fifth taste—umami (savory) has become widely accepted, and several other tastes (fattiness, hotness) have been suggested. The notion of a tongue or taste map, where taste buds on certain parts of the tongue are responsible for sensing certain tastes, has also been disproven. In fact, all areas of the tongue contain receptors for all of the tastes.

Some things can make tastes less pronounced. Temperature has an impact on the sensitivity of taste cells. Cold food and drinks reduce taste sensitivity. That's why frozen juice pops never taste as sweet as the juice used to make them. Smell also has a profound affect on taste. If you have a cold and your nose is stuffed up, food will not taste as strong as it otherwise would because your sense of smell is impaired.

Fingernails Are Crunchy

If I could think of all the things
I ate since I was small.
I'd have to have a ream of paper
Just to list them all.

Well, fingernails are crunchy,
But I guess that's not a taste.
I've sampled dirt and crackers
Then let's see—there's white school paste.

And surely it makes cookies good
Though how is hard to tell,
For I've tried vanilla extract
And it's nothing like its smell!

Once I spent the night at grandma's
And 'till now she doubts my word;
That she left it, not the toothpaste,
By my toothbrush seems absurd.

But would I, and all so vividly,
If this had been a dream,
Recall 'till now the awful taste
Of her hair-styling cream?

Another time, out on the porch
My mother came to see,
Since I was sent to feed the pets,
Just what was keeping me.

And there was I, cats all around,
Cats on the porch and underneath,
The can of cat food in my hand,
The spoon between my teeth.

I think it safe for me to say
That sour or salty as the sea,
Of all the tastes there are to taste,
I've tasted a variety.

Brenda Dahl

What tastes good to you?
Can you give us just a clue?
Is it cocoa that you drink
When the day is through?
Is it ice cream, cool and sweet,
Or the apples that you eat?
What tastes good to you?

For your tongue can tell,
And it does it very well:
Special taste buds give you taste,
Helped by what you smell.
Salt and sour and sweet
And the bitter things you eat—
Yes, your tongue can tell.

SEEING IS NOT ALWAYS BELIEVING

Topic
Sense of taste

Key Question
What senses do we use to tell how substances are different when they all look the same?

Learning Goal
Students will begin to understand that their sense of sight is not always reliable when predicting how something will taste.

Guiding Documents
Project 2061 Benchmark
- *People use their senses to find out about their surroundings and themselves. Different senses give different information. Sometimes a person can get different information about the same thing by moving closer to it or further away from it.*

*NCTM Standard 2000**
- *Represent data using concrete objects, pictures, and graphs*

Math
Data organization
  Venn diagrams

Science
Life science
  human body
    senses

Integrated Processes
Observing
Comparing and contrasting
Predicting
Classifying
Applying

Materials
For the class:
  4 clear zipper-type plastic bags
  ½ cup salt
  ½ cup flour
  ½ cup powdered sugar
  ½ cup granulated sugar
  water
  bulletin board paper (see *Management 5)*
  Venn diagram(see *Management 6)*

For each student:
  hand lens
  plastic spoon
  paper plate (see *Management 3)*
  Recording Sections
  3-oz paper cup
  scissors
  glue
  crayons

Background Information
The taste buds on our tongues contain sensory receptors that provide us with information about how things taste. The sense of sight helps us to anticipate the taste of foods. When the substances look very similar, it is often a surprise to find that they actually taste different. This activity has students trying to identify the tastes of substances that are very similar in appearance.

Management
1. This activity is best done in small group settings. Learning stations are highly recommended.
2. An adult supervisor is essential in ensuring that this activity is successful.
3. Prior to the activity, use a marker to divide paper plates into four sections. Number the sections one to four.
4. Copy one *Recording Sections* page per student.
5. Prepare a class chart by dividing bulletin board paper into four sections. This should be large enough to accommodate the cut quarters of students' paper plates (see *Procedure 10)*.

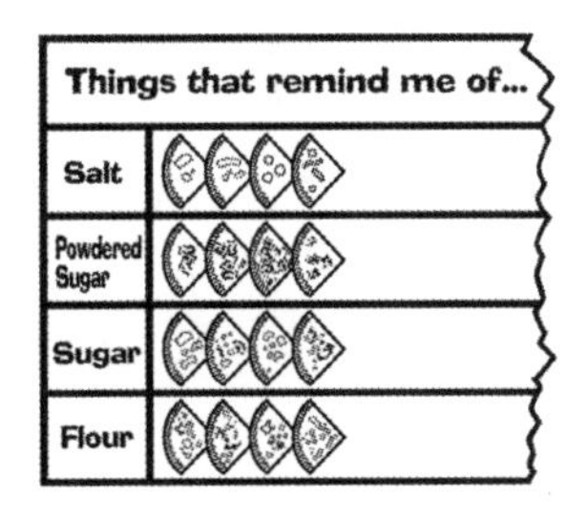

6. Enlarge the two-circle Venn diagram with the labels: sweet, salty.

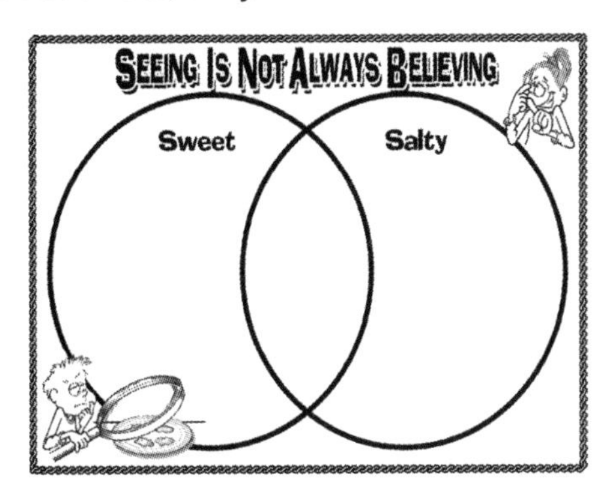

7. Number the four plastic bags. Put ½ cup salt into bag #1, ½ cup powdered sugar into bag #2, ½ cup flour into bag #3, and ½ cup of sugar into bag #4.
8. Hand lenses are available from AIMS (item number 1977).

Procedure

1. Allow the students to look at the four plastic bags containing samples of each of the substances you have prepared for them to taste. Ask them if they can tell just by looking at the substances what each one is. Allow the students to give suggestions as to what each substance might be.
2. Make a temporary class graph on the board to record their predictions as to what each bag might contain.
3. Ask the *Key Question* and discuss the dangers of tasting unknown substances without adult supervision. Reassure students that these substances are not poisonous.
4. Give students plastic spoons, the quartered paper plates, and a small cup of water. Direct them to put approximately ¼-½ teaspoon of the first substance (salt) onto their plates making certain that substance #1 goes onto section #1 of their paper plates.
5. Before the students taste the first substance, give each student a hand lens to look closely at it. Discuss what it looks like (the crystal shape), what it smells like, etc.
6. Ask the students if they want to change their original predictions of what they think the substance might be. Let them taste the substance.

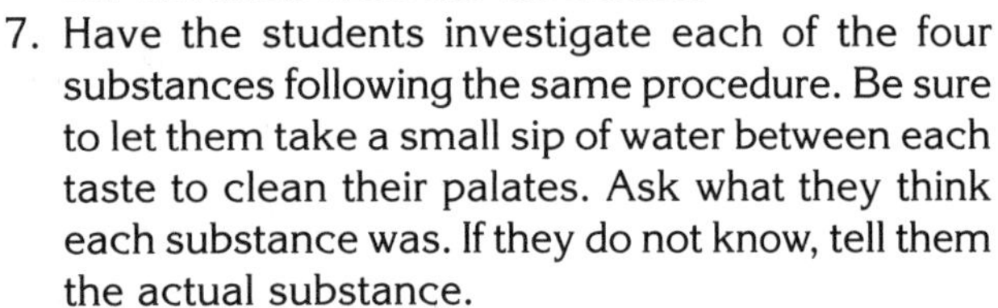

7. Have the students investigate each of the four substances following the same procedure. Be sure to let them take a small sip of water between each taste to clean their palates. Ask what they think each substance was. If they do not know, tell them the actual substance.
8. If any substances are left on the plates, have students shake them off into the trashcan. Direct students to draw foods that have the taste of the mystery substance in the appropriate section of the paper plate. (Examples: chips, candy, pancakes, frosting) You may want to label the drawings.
9. Have students cut their paper plates into the four sections and glue them onto the chart (see *Management 5).*
10. Distribute the *Recording Sections* page to each student. Have the students number the sections from one to four. Ask the students to draw something (or paste in pictures from old magazines) in the numbered sections. Section #1 should have a picture of something salty; #2, something made with powdered sugar; #3, something made with flour; and #4, something made with sugar. Have the students cut these sections out and use them on a class Venn diagram (see *Management 6).*

Connecting Learning

1. When we looked at these substances and noticed that they looked a lot alike, what sense did we decide to use to tell if they were the same or different?
2. Let's look at each of the numbered sections on our class chart and see what we thought each of the substances tasted like. (Go through each of the chart sections.)
3. What did you use to find out what these things tasted like? [sense of taste]
4. What part of our body contains the sense of taste? [mouth, tongue]
5. How does your tongue help you taste? (Discuss taste buds.)
6. Where are your taste buds located? (If you have a mirror, you may want students to observe the taste buds on their tongues.)
7. Why should we not taste unknown things when there are no adults around to supervise?
8. Do some foods taste both sweet and salty? Let's look at our Venn diagram and see which ones taste just sweet. ...just salty. ...both. Do we all agree about the tastes? Why not?
9. What are you wondering now?

Extensions

1. Do the same activity with liquids: water; flat, clear soda; clear gelatin dissolved in water, white vinegar, sugar water.
2. Do the same activity with peeled fruits and vegetables: apples, potatoes, pears.
3. Using wrappers from the students' lunches (or pictures of food items), graph them according to the attributes of salty, sweet, and sour.

Home Links

1. Ask students to find things that are salty or sweet at home. Draw pictures to represent these items.
2. Ask students to keep a record of the number of foods that are salty or sweet that they ate for breakfast or lunch or dinner.

Seeing Is Not Always Believing
Sweet
Salty

Seeing Is Not Always Believing

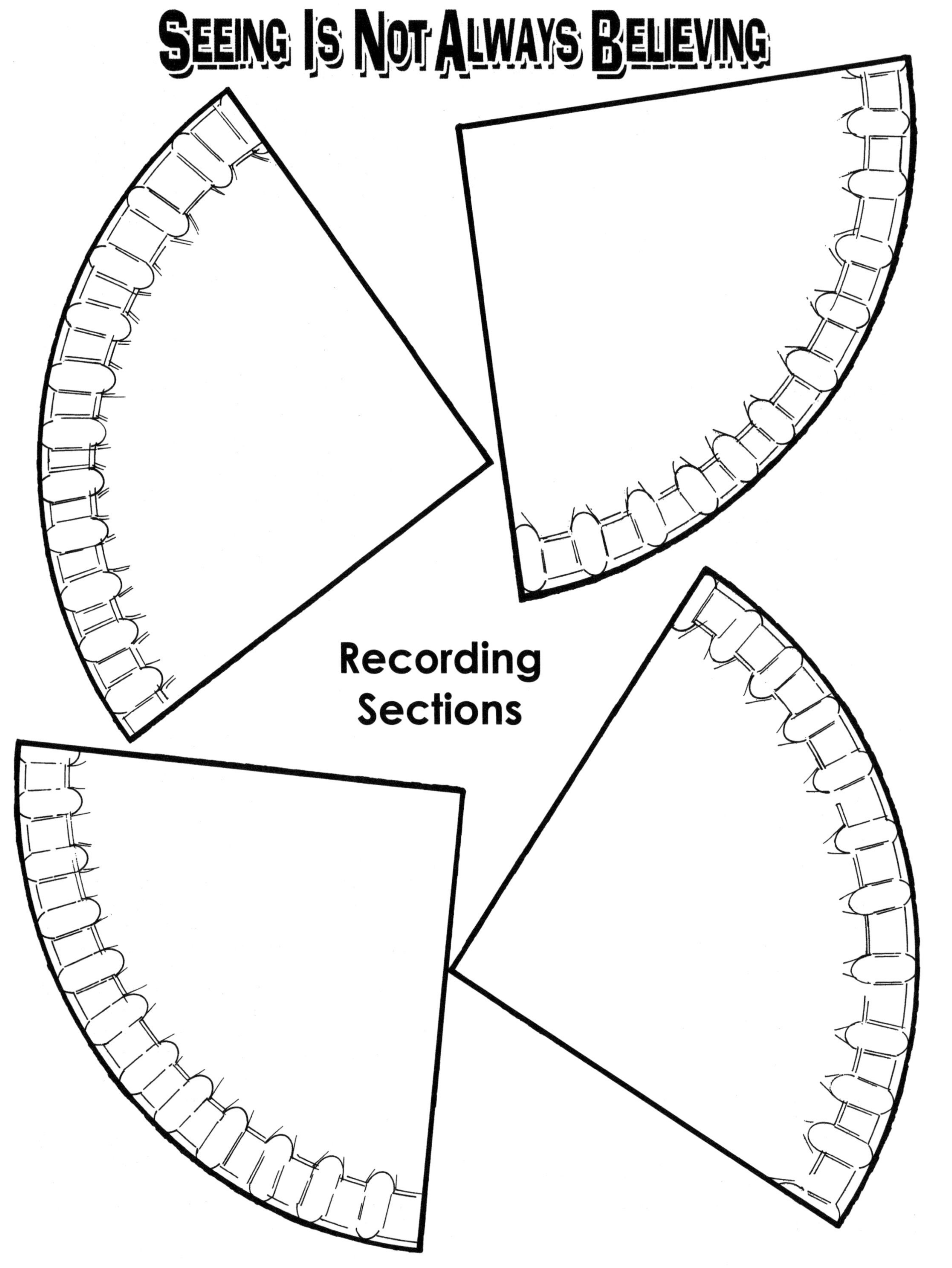

EGGS-TRA SPECIAL SCRAMBLE

Topic
Senses of taste and sight

Key Question
How will the color of the eggs affect how they taste?

Learning Goal
Students will become aware of how the sense of taste is affected by the sense of sight.

Guiding Documents
Project 2061 Benchmark
- *People use their senses to find out about their surroundings and themselves. Different senses give different information. Sometimes a person can get different information about the same thing by moving closer to it or further away from it.*

*NCTM Standard 2000**
- *Represent data using concrete objects, pictures, and graphs*

Math
Graphing

Science
Life science
  human body
    senses

Integrated Processes
Observing
Comparing and contrasting
Collecting and recording data
Interpreting data
Predicting

Materials
For the class:
  Green Eggs and Ham by Dr. Seuss
  2 class graphs
  glue sticks
  1-2 dozen eggs (see *Management 2)*
  1 or 2 electric skillets or hot plate and skillets
  butter or non-stick cooking spray
  green food coloring
  milk
  salt and pepper
  spatula

For each student:
  graph markers (see *Management 2)*
  paper plate
  fork

Background Information

The tongue sends messages to the brain about what we eat. The tongue is covered with taste buds that help us to taste things. The tongue, however, is not the only important sense in tasting. The senses of smell and sight also play important roles. If you closed your eyes and held your nose, you might not be able to tell the difference between an apple and a potato. The message your sense of sight sends the brain before you even taste the food often plays an important role in determining likes and dislikes.

Management
1. Prepare two class graphs by enlarging them. One will be used for predictions and one for actual results.
2. Cut out the graph markers; each student will need two markers.

3. Set up cooking area. Another adult could cook one batch of eggs if two skillets are available or simply do the cooking procedure twice.
4. The number of eggs needed is dependent upon the number of students.

Procedure

1. Read the book *Green Eggs and Ham* by Dr. Seuss.
2. Ask how many students have eaten green eggs. How do you think they would taste? Would they taste the same as yellow eggs?
3. Ask how we could determine if they taste the same.
4. Show students one graph and explain that they will use it to record their predictions about how they think the eggs will taste.
5. Distribute one graph marker to each student. Instruct students to write their initials on their markers. Have them glue their markers in the appropriate section of the prediction graph.
6. Make scrambled eggs by beating eggs, milk, salt, and pepper in a bowl. Divide the mixture in half. Add green food coloring to one of the egg mixtures and cook in buttered or sprayed non-stick pans. Leave the other egg mixture as is (yellow).
7. Give each student a small portion of each type of egg on the paper plate.
8. Invite them to taste the green eggs and discuss how they taste.
9. Have them taste the yellow eggs. Discuss.
10. Distribute a second graph marker to each student for recording on the second graph representing the actual taste. Let students initial and glue their markers in the section they choose.
11. Compare the two graphs.

Connecting Learning

1. How are the two graphs different?
2. How close were your predictions to the final results?
3. Does anybody know what was added to change the color? Does food coloring have a flavor? If it does not have a flavor, could it change the flavor of food?
4. How did our eyes change how we felt about the food? How did our eyes change the taste of the food?
5. What are you wondering now?

Extensions

1. Taste white and dark chocolate. Discuss their similarities and differences.
2. Reread the book, *Green Eggs and Ham,* and discuss how the animal in the book changed his mind about green eggs. What made him change his mind?
3. Discuss or show other foods that are different colors than expected, such as red pears or broccoflower.

Curriculum Correlation

Language Arts

Using the framework in the Dr. Seuss book, write a poem about a funny-colored food. Let students draw pictures of foods that are different colors and share their work.

Home Link

Send home the recipe for eggs and ask if anyone wants to try other colors of eggs and report to the class.

EGGS-TRA SPECIAL SCRAMBLE

Do green and yellow eggs taste the same?

Same **Different**

EGGS-TRA SPECIAL SCRAMBLE

Copy and cut apart the graphing markers so that each student can have two.

Topic
Sense of taste

Key Question
How do you decide if you like one kind of food better than another?

Learning Goal
Students will compare the taste of three kinds of porridge.

Guiding Documents
Project 2061 Benchmark

- *People use their senses to find out about their surroundings and themselves. Different senses give different information. Sometimes a person can get different information about the same thing by moving closer to it or further away from it.*

*NCTM Standard 2000**

- *Represent data using concrete objects, pictures, and graphs*

Math
Data organization
  Venn diagrams
  graphing

Science
Life science
  human body
    senses

Integrated Processes
Observing
Comparing and contrasting
Collecting and recording data
Interpreting data
Inferring

Materials
For the class:
  Goldilocks and the Three Bears
  three types of hot cereal *(see Management 2)*
  salt, optional
  margarine, optional
  water
  hot plate or microwave
  cooking pot or bowl
  wooden spoon
  measuring spoons
  measuring cups
  Yum and Yuk chart, enlarged
  Yum and Yuk Venn diagram
  My Favorite Porridge graph

For each student:
  1 plastic spoon
  1 small paper plate
  1 set of graphing markers
  1 student book

Background Information
Students need many opportunities to explore food and cooking. In this activity, students will explore the taste and texture of food. They will focus on the "cooked" texture as they taste and compare three kinds of cereal and choose their favorite and least liked. They will observe the changes brought about by the addition of water and heat to the dry cereals.

Management
1. Prior to class, enlarge the class graph, the *Yum and Yuk* chart, and the Venn diagram. You may choose to use one, two, or all three representations of data; the variety is provided for your convenience.
2. Use three different types of hot cereal such as oatmeal, cornmeal, and a creamy wheat cereal.
3. Use the recipe guides on the boxes of cereal for preparation. Prepare a quantity according to the number of students in your class. Each student will probably need about one-fourth of a normal serving.
4. This activity can be done in a variety of ways depending on your adult-to-student ratio.
   - Divide the class into groups, each preparing one of the cereals and coming together to taste, compare, and discuss the textures.
   - Teacher demonstrates the preparation of the cereals to the class involving the students as helpers, then they proceed with comparing taste and texture.
   - Reduce the recipe into individual portions and allow each student to make his or her own cereal. **Caution: You will be using very hot water in the preparation of these cereals.**
5. A set of graphing markers for each student is necessary if you are using all three representations of data. A set of markers consists of:
   - three *Yum* and three *Yuk* markers for the *Yum and Yuk Chart.*
   - two graphing markers with students' names: one for the Venn diagram and another for the *My Favorite Porridge* graph.

Procedure

1. Ask students if they have heard the story of *Goldilocks and the Three Bears.* Ask them to briefly recall the storyline. You may want to read one or more versions of the story at this time (see *Curriculum Correlation*).
2. Direct the students to think about which senses the characters are using in the story.
3. Discuss what porridge is and what kind the bears may have left to cool.
4. Ask the students how they could find out whether they like porridge or not. [taste some]
5. Have students observe and describe the cereal before it is prepared.
6. Prepare the cereals by following package directions. Let students help by allowing them to do any of the following: measure, pour, and stir (before temperature gets to the boiling stage). **Caution: Hot cereals bubble when boiling. Have students maintain a safe distance from boiling cereal.**
7. Let the cereals cool.
8. Place a teaspoonful of one of the cereals on the students' tasting plates.
9. Discuss how the cereal looks after it has been cooked.
10. Tell students the name of the cereal and have them taste it.
11. Have students color and cut appropriate *Yum /Yuk Markers* and place them on the *Yum and Yuk Chart.*
12. Follow the same procedure for the other two cereals.
13. Have students cut out and write their names on two graphing markers. Direct them to place their markers in the appropriate column on the *My Favorite Porridge* graph and in the appropriate area of the Venn diagram.

Connecting Learning

1. How did the cereal change after it was cooked? Does there seem to be more cereal after it is cooked? Why do you think this happens?
2. Will cooking all kinds of cereal do the same thing that happened to these cereals? Explain your answer.
3. What do you think would happen if we cooked the cold cereal some of you eat in the morning?
4. What made our cereal change? [water and heat]
5. If we were going to do this again, what cereal might we use?
6. Compare the class results from the chart, graph, and Venn diagram.
7. Did everyone in our class like the same cereal? Why or why not?
8. Is one cereal better than another because more students like it? Explain.
9. By looking at the results on our chart and graph, if we were going to make a breakfast for this class, would we need to make all three cereals? Explain.
10. What are you wondering now?

Extensions

1. Talk about the other senses used by the story characters in various scenes.
2. Have each student make a copy of the class graph by coloring pictures in each column to represent the number of likes and dislikes for each of the three cereals.

Curriculum Correlation

Brett, Jan. *Goldilocks and the Three Bears.* Paperstar. New York. 1987.

Marshall, James. *Goldilocks and the Three Bears.* Dial Books. New York. 1988.

Turkle, Brinton. *Deep in the Forest.* Dutton Children Books. New York. 1976.

Home Links

1. Have a taste test with the family.
2. Graph the results of your family's most and least favorite cereals.
3. Help cook your favorite cereal for the family.
4. Watch a parent cook your favorite cereal to see if he/she follows the same steps.
5. Bring in your favorite cereal for a tasting party.
6. Make a list of food you could cook for breakfast.

Copy and cut apart these markers to use on the Yum and Yuck Chart. Each student will need three markers.

Make Mine Porridge

Yum and Yuk Chart

Cornmeal		Oatmeal		Creamy Wheat	
yum	yuk	yum	yuk	yum	yuk

Color and cut apart the graphing markers for use on the Venn diagram and the *My Favorite Porridge* graph. Each student needs one.

Make Mine Porridge
Oatmeal
CORN MEAL
Creamy Wheat
Hot Cereal

My Favorite Porridge

Cornmeal	Oatmeal	Creamy Wheat

Make Mine Porridge

Three bears: small, medium, and large,
are planning for a feast,
so they must choose their favorite dish,
and which they like least.

They start the list with porridges:
creamed wheat, cornmeal, and oat.
They sample each, then once again—
and now its time to vote.

TASTE TEST

Topic
Sense of taste

Key Question
What body part helps us taste things?

Learning Goal
Students will see if they can distinguish the tastes of apple pieces sprinkled with salt, sugar, and lemon juice.

Guiding Document
Project 2061 Benchmark
- *People use their senses to find out about their surroundings and themselves. Different senses give different information. Sometimes a person can get different information about the same thing by moving closer to it or further away from it.*

Science
Life science
  human body
    senses

Integrated Processes
Observing
Comparing and contrasting
Communicating
Applying

Materials
Apples (see *Management 1)*
Chart (see *Management 3)*
Plastic food service gloves
Salt
Sugar
Lemon juice
Paper plates

Management
1. Prepare the apples ahead of time. Peel and cut apples into bite-sized pieces. (Peeling the apples ensures that the skin will not influence students' perceptions of taste.) Each student needs three pieces of apple. Dip a third of the apple pieces in lemon juice. Sprinkle another third with salt and the remaining third with sugar.
2. Divide one paper plate per child into three sections. Label these *A*, *B*, and *C*. Place one piece of the sugar-sprinkled apple in section *A*. Place one piece of the lemon-dipped apple in section *B*, and place one piece of the salt-sprinkled apple in section *C*.
3. Prior to teaching the lesson, copy the pictures of the students to make a three-column chart. Glue the words *sour*, *sweet*, and *salty* along the left side of the chart and place the student pictures at the top as illustrated below.

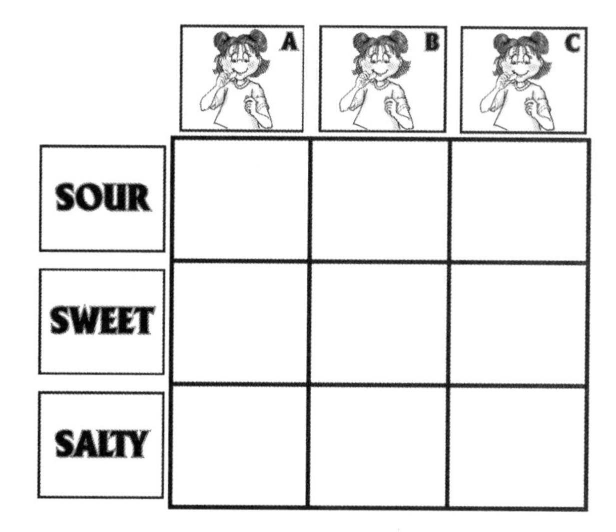

Procedure
1. Discuss various tastes such as salty, sweet, sour, etc. Encourage the students to name several examples of each taste. Record student responses on the board. Question students about what body part or parts we use to taste our food.
2. Explain to the students that they will be participating in a tasting experiment. Tell them that they will be given three different types of apples—salty, sweet, and sour. Give each student a sample plate.
3. Have students eat one of the apples from section *A*.
4. Ask them to raise their hands if they think the apple tasted *sour*. Record the corresponding number of tallies in the appropriate section of the chart. Ask the students to raise their hands if they think the apple tasted *sweet* in their mouths. Record the number using tally marks next to the word *sweet*. Ask them if they think the apple tasted *salty* in their mouths. Record the number using tally marks next to the word *salty*.
5. Repeat this procedure, having students test the remaining two samples.

6. Draw students' attention to the tally chart and discuss where the most correct tallies are located for each taste test.
7. Discuss the actual flavorings for the apples.
8. End with a discussion about the senses and related body parts that are involved in tasting our food.

Connecting Learning

1. Were you able to tell the flavor of the apples you were eating?
2. Which flavor (sour, sweet, or salty) was easiest to taste? ...hardest?
3. What body part(s) helped you know what you were eating??
4. What are you wondering now?

TASTE TEST

SOUR	
SWEET	
SALTY	

Smell

What smells better than to walk into your house and smell freshly baked bread? Maybe it's the smell of a clean baby or your favorite cologne or the outdoor smell at a summer cabin. Smells are powerful forces for triggering memories.

The sense of smell, or olfaction, originates in your nose. When you inhale, you can feel the air being drawn into your nose. This air is filled with odor molecules that flow through your nasal cavity where they encounter hair-like projections that are covered with a layer of mucus. The mucus dissolves the odor molecules so they can be sampled by the receptors. Because you are always breathing, you are always monitoring the odors in the air. When you want to smell something in more detail, all you have to do is take a good sniff to bring more air into your nasal cavity. This causes more odor molecules to pass over the olfactory hairs helping you to better perceive the smell.

Sometimes the sense of smell gets desensitized. If you walk into a room where a scented candle has been burning, you might notice the odor at once. Yet, after a while, you no longer notice it. The sense of smell responds mainly to changes in stimuli. Once you become accustomed, or desensitized, to the smell, it may have to increase up to 300 times in strength before you notice it again. This helps prevent you from becoming overloaded with unimportant information and is a good example of how your brain acts as a filter of information. The people who live in the house may be so used to the smell of the scented candle that they do not notice it. However, if they have a natural gas leak, they would notice this odor even though they were not conscious of the scent of the candle. The natural gas leak represents a change in the smells to which they were accustomed.

The sense of taste and the sense of smell are very closely linked. When you eat food, it gives off odor molecules that travel up the connecting passage between your mouth and your nose. The odor molecules enter the nasal cavity stimulating the smell receptors. The food's flavor is a result not only of what is tasted but also what is smelled. You've surely noticed that when you have a cold, your food doesn't seem to have as much flavor. This is because your nose is blocked by thick mucus that doesn't allow the odor molecules to penetrate.

Compared to other animals, you do not have an especially good sense of smell. Dogs are often trained to track people or sniff out certain substances such as drugs. They are able to do this because they may have 100 million olfactory receptors compared to your five million!

Especially for You

My mom is not at home today.
The rain outside is splashing down,
And though I can't go out to play,
I love the Earth smells all around.

My grandma's come to stay with me.
My grandpa's here as well.
My house is filled up pleasingly
With every wondrous smell.

There's onion, roast, and garlic,
Warm yeasty breads of wheat and rye,
Ginger, cloves, burned candlewick,
And wet galoshes left to dry.

A thunder clap! And like a flash,
Throwing wide the kitchen door,
In slides my Irish Setter, Splash,
'Cross grandma's new-waxed kitchen floor.

My grandma isn't smiling—yet.
Her nose is wrinkled up.
"Oh, Splash! You smell just horrid wet!"
She scolds my dripping pup.

Yet 'mid all this confusion, a warm familiar smell,
Drifts in among the others, and simply lingers there.
I turn, and in the doorway, I'd heard no knock or bell,
My mom and dad just watch and smile, while I just stand and stare.

My mom kneels down upon the rug, so I sit down there too.
She says, "I've brought home something, just especially for you."
The something that she hands me is all wriggly, wrapped in blue.
It's the tiniest baby sister, when I kiss her she smells—New!

Brenda Dahl

My Sense of Smell
Words by Suzy Gazlay
Tune: Animal Fair
I have a sense of smell and it helps me to
tell when food tastes good the way it should; it
does its job quite well. I smell a sum-mer
breeze; some-times it makes me sneeze. My
nose can tell the won-der-ful smell of
flow-ers, grass, and trees.
Now what can that smell be?
I smell but I can't see
My mom's perfume
Across the room
It sure smells good to me!
I love to smell the sea,
A piney Christmas tree,
I'd love to tell
My favorite smell
Won't you tell yours to me?
© Suzy Gazlay 2003. Used by permission.

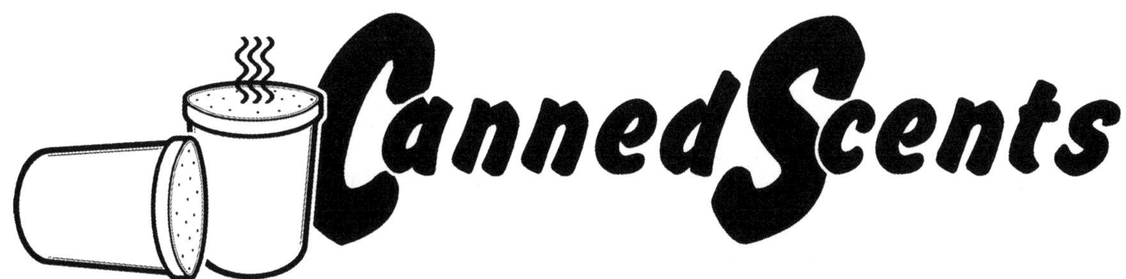

Canned Scents

Topic
Sense of smell

Key Questions
1. Do you think you can identify what an object is by just smelling it?
2. What do some smells remind you of?

Learning Goal
Student will become aware of the information given to them through the sense of smell.

Guiding Document
Project 2061 Benchmark
- *People use their senses to find out about their surroundings and themselves. Different senses give different information. Sometimes a person can get different information about the same thing by moving closer to it or further away from it*

Science
Life science
human body
senses

Integrated Processes
Observing
Classifying
Predicting
Comparing and contrasting
Collecting and recording data
Interpreting data

Materials
For the class:
Our Smell Chart (see *Management 4)*

For a station:
4 empty 35 mm opaque film canisters with lids
diced onion pieces
orange sections
3 tablespoons ground coffee
bar of soap cut into small pieces
scissors
tape
crayons

For each child:
student recording sheet
clipboard
crayons and/or pencils
My Home Smells Nice student booklet (see *Management 5)*

Background Information
There are about five million olfactory receptors in the human nose. These receptors give humans the ability to discriminate approximately 10,000 different odors. Even though this seems like an extraordinary feat, our olfactory abilities compare very poorly to those of other mammals. The sense of smell guides animal behavior more than any other sense. Dogs use their noses to explore things around them in the same way we humans use our eyes. A dog's olfactory sense is many, many times more sensitive than ours. Despite this great difference, humans find smells to be distinctive, identifiable, and memorable.

Management
1. Poke a few small holes in the top of each film canister lid. If film canisters are not available, other small opaque containers with lids can be substituted.

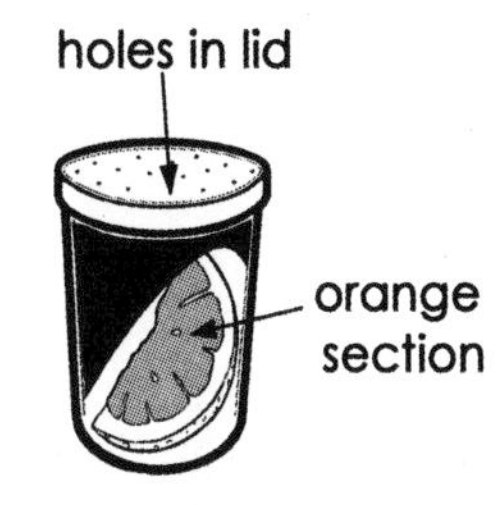

2. Place the onion in one film canister, the orange into another, coffee into a third, and the soap bits into the last. Number the film canisters using a permanent marker.
3. At each station you will need to provide the following: the four filled and numbered film canisters, student recording sheets, crayons, and scissors.
4. Enlarge the *Our Smell Chart* to record the students' predictions.
5. Prepare a *My Home Smells Nice* booklet for each child. Cut the page into fourths and staple on the left side.
6. This activity is best taught in two parts. The first part involves students in small groups determining and charting the mystery contents of film canisters. The second part asks students to relate scents to different memories—walking or riding to school, home scents, etc.

Procedure

Part One

1. Hand out the student recording sheets with the pictures of numbered film canisters. Have students attach sheets to their clipboards to provide a smooth writing surface.
2. Show students how to fan the air above the canister toward their noses in order to smell the contents. After allowing the students to smell each canister, have them draw their predictions of the contents in the appropriately numbered canister on the recording sheet.
3. When the group is finished smelling the canisters, have the students cut on the lines of their papers and place the numbered drawings in the appropriate columns on the enlarged *Our Smell Chart.*
4. Invite other groups to the station until all class members have made their predictions and contributed to the chart.
5. Share the finished chart with the entire class.
6. Identify the actual contents of each canister and discuss the results.

Part Two

1. Fold a piece of 8.5 x 11-inch paper in fourths. Number the sections of the paper 1 through 4. Have students attach the paper to their personal clipboards.
2. Take the students on a walk to four specific areas. Some suggested areas are behind the cafeteria near the garbage bin (yuk!), out on the playground just after the grass has been mowed, near the cafeteria while lunch is cooking, and in the parking lot. Instruct them to carefully observe the smells. Have them record (by drawing or writing) what they smell.
3. Return to the classroom and talk about the smells the students noticed, and the ones they recorded (see *Connecting Learning).*
4. Ask the students who walk to school what pleasant smells they noticed along the way. Contrast this with the smells noticed by the students who came to school on the bus or in a car.
5. Point out that different smells remind us of different places.
6. Discuss with the students how smells can remind us of things or events that happened in the past. You might want to bring in a scent of an evergreen tree, spray it and allow the students to smell it and ask them what it reminds them of. [Christmas] Other scents (potpourri) you might bring would be vanilla or cinnamon to remind students of things baking at home.
7. Distribute the *My Home Smells Nice* student booklets and have the students draw or write about smells they can remember from home or assign this page for homework to discuss the following day.

Connecting Learning

Part One

1. What did your nose tell you about what was in canister one? ...two? ...three? ...four?
2. Did everyone smell the same thing in canister one? ...two? ...three? ...four? How can you tell? [by looking at the pictures on the chart]
3. Which scent did the most students identify correctly? ...the fewest students?
4. What does your nose tell you about what was in the containers that your eyes did not?
5. When you smelled the contents of the containers, did any memories or places come to mind? Explain.
6. How are your memories or places that the smells make you think of like those of your classmates? How are they different?

Part Two

1. What kinds of smells did you experience on your walk? [good, bad, earthy, wet, etc.]
2. Did everyone have the same reaction to all of the smells? What are some smells that you liked/disliked that a classmate disliked/liked?
3. What smells did you experience on the way to school? Were the smells different for those who walked compared to those who rode in a bus or car? How were they different? How were they the same?
4. What are some smells that you experience at home? How do they compare to the smells your classmates experience at home?
5. Where do you experience the most pleasant smells? Why?
6. What are you wondering now?

Canned Scents

1
2
3
4

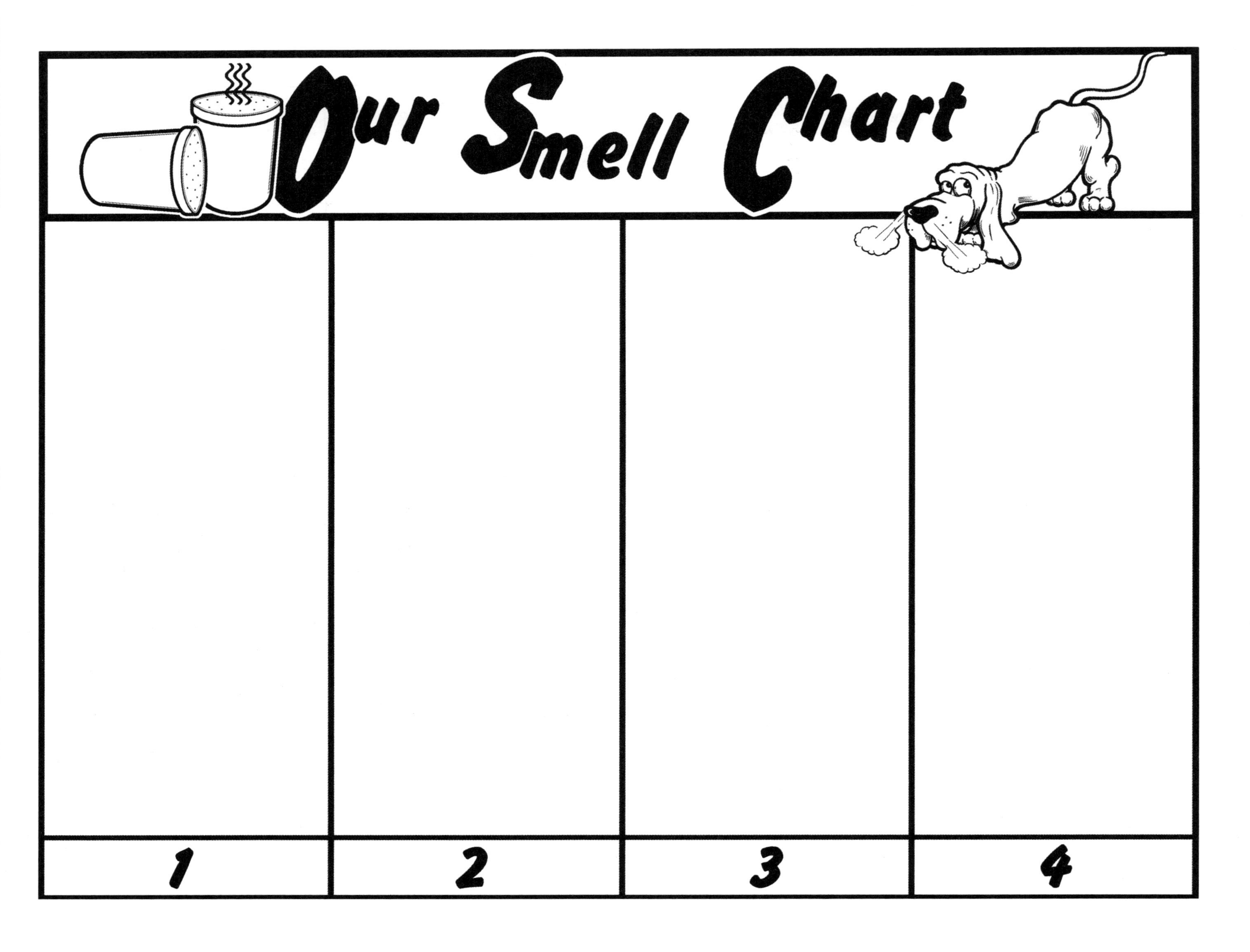
Our Smell Chart
1
2
3
4

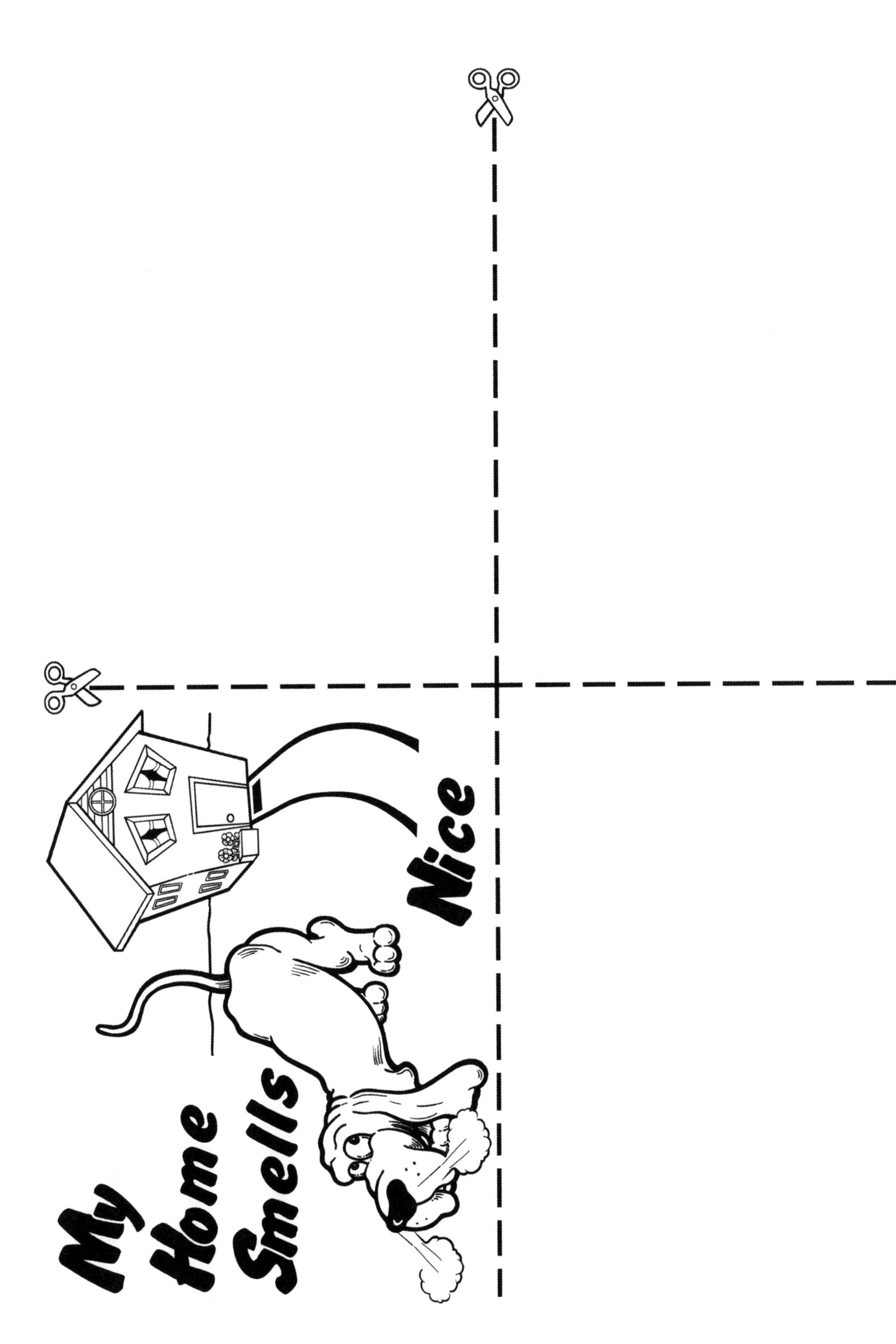
My Home Smells Nice

The Napping Nose

Topic
Sense of smell

Key Question
How long do you think you can smell the potpourri?

Learning Goal
Students will become aware of how the sense of smell becomes less sensitive to a scent with the passage of time.

Guiding Documents
Project 2061 Benchmark
- *People use their senses to find out about their surroundings and themselves. Different senses give different information. Sometimes a person can get different information about the same thing by moving closer to it or further away from it.*

*NCTM Standard 2000**
- *Represent data using concrete objects, pictures, and graphs*

Math
Graphing

Science
Life science
  human body
    senses

Integrated Processes
Observing
Comparing and contrasting
Predicting
Collecting and recording data
Interpreting data

Materials
Potpourri
Hot plate and pan or electric hot pot
Red construction paper (see *Management 1)*
Blue construction paper (see *Management 2)*
Glue or tape
Class graph (see *Management 5)*
Clock
Safety pins, one per student

Background Information
The nose smells through the use of the olfactory membrane, which is located in the nasal passage just behind the bridge of the nose. This membrane is made up of sensory cells that send messages about scents via the olfactory nerve to the brain. The sensory cells in the olfactory membrane may become desensitized to the scents to which they are exposed over long periods of time. When you first walk into a room, you may smell many things: dinner cooking, perfume, flowers, smoke; but, after you have been in the room awhile, the olfactory nerves "get tired" and you no longer smell these odors.

Management
1. Prior to this activity, cut up ¾" x ¾" squares of red construction paper. You will need eight squares per student. The red construction paper is used to indicate that students can smell the potpourri.
2. Cut blue construction paper into ¾" x ¾" squares. You need one square per student. The blue construction paper is used to indicate that students cannot smell the potpourri.
3. Sticky notes may be used as an alternative to the cut-up squares of construction paper. The sticky notes need to be cut to fit the chart.
4. Prepare the class chart by copying the provided sheets onto blue paper. There is room for 22 students to record their data. If you have more than 22 students, make an extra copy of the second page. Any extra squares should be cut off so the chart is one square along its horizontal axis for each student. Glue or tape the necessary sheets together. (If you want to enlarge the chart, you will also need to enlarge the construction paper squares to fit.)
5. Prepare the chart as follows:
   a. number the bottom of the chart according to the number of students particiating in the activity
   b. draw in the hands on the clocks at 10-minute intervals
6. This activity takes one hour to complete. It is suggested that you plan either station work or group work that can be easily interrupted at 10-minute intervals (free exploration, manipulative discovery time, reading stories or books, show and tell, etc.) for the hour.
7. Before school begins, put some potpourri in a hot pot and heat the pot to release the potpourri's scent.

Procedure

1. Have the whole class sit in a circle and discuss what is different in the room. [smell] Ask the *Key Question*: How long do you think you will be able to smell the potpourri?
2. Explain to the students what you have in the hot pot. Ask every student who can smell the scent to stand. For students who can smell the potpourri, pin a red square on their shirts and have them glue or tape a second red square on the bottom line of the class chart starting on the left side continuing to the right. If students cannot smell the potpourri, pin a blue square on them. These students will not need to place a square on the chart as the chart will show a blue space because of the lack of the red square.
3. Compare the amounts of blue spaces and red squares on the bottom line of the chart. (If all the students can detect the scent, the entire bottom line of the chart will be red squares with no blue chart paper showing.)
4. Tell the students to compare the red squares they are wearing to the red squares on the graph. Explain to them that the graph on the wall is a representation of our class. If they are all wearing red, then the graph at the *0 minutes* line should all be red. But if anyone can't smell the scent, there will not be a red square to represent that person. The graph will show blue space for that student.
5. At 10-minute intervals, ask the students whether or not they can smell the potpourri. Those who can still smell it should put red squares on the appropriate line of the graph, always starting from the left and working to the right. Those who can't should exchange the red squares that are pinned on their shirts for blue squares. The students who do not smell the scent will add nothing to the graph because the blue automatically shows on the chart. Compare the amounts of red and blue showing on the graph with the amounts of blue and red squares on the students' shirts.
6. At the end of an hour, ask the students to sit in a large circle. Discuss how the number of people who could smell the potpourri changed and how the chart has comparatively changed.

Connecting Learning

1. At the start of class, how many students put a red square on the *0 minute* line of the graph? How many students wore a red square at that time? What did that tell us? [Everyone who could smell the potpourri wore a red square.]
2. As the time passed, what happened to the color on our graphs and the squares on our shirts? [more became blue]
3. What does this tell us about the number of students who smelled the scent?
4. Have the students go outside the room for a short walk or recess. Upon coming back into the room, ask the students if they notice anything about the scent of the room. Count the number of students who can now smell the potpourri. Why do you think more of us can now smell the potpourri?
5. Why could we no longer smell the scent when it was still there? [We got used to it; our noses got "tired."] Explain that sometimes our olfactory sense gets desensitized, or "tired," and then we cannot smell something we smelled before.
6. How do smells keep us safe? [We can smell smoke that warns us of fire. We know by smell that there may be a gas leak. We can identify gasoline or rubbing alcohol by their smells and know that they are dangerous to us.]
7. What are you wondering now?

Extension

Visit the cafeteria near lunch time and note the different smells. Have students think about how long they can smell the cooking smells in the cafeteria when they go to lunch.

Home Link

Have students draw what they smell when they walk into their home after school. Ask Mom or Dad to help keep track (in minutes) of how long they can smell it.

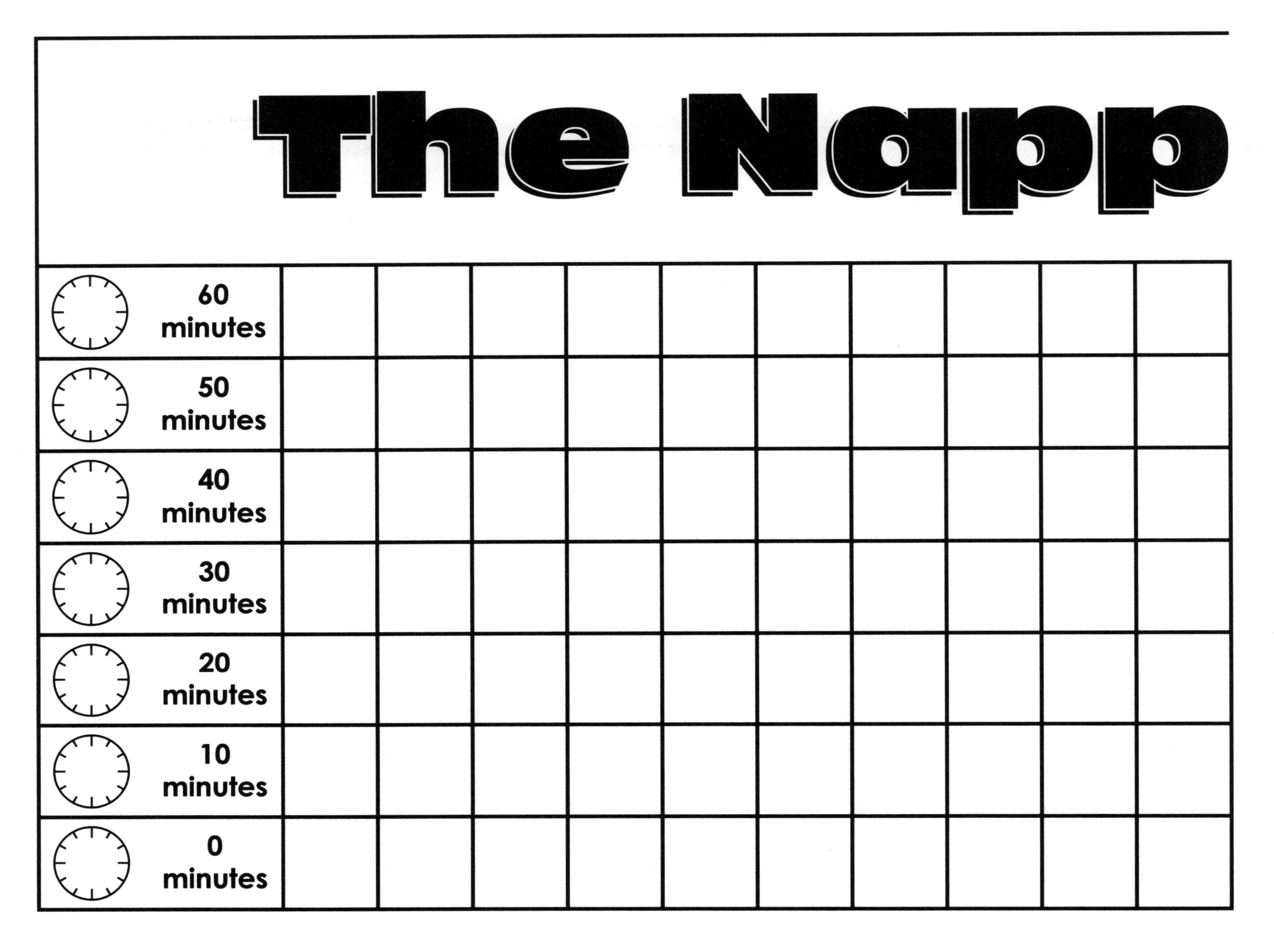
The Napp
60 minutes
50 minutes
40 minutes
30 minutes
20 minutes
10 minutes
0 minutes

ing Nose

Making Sense of What You Smell

Topic
Sense of smell

Key Question
What sense can we use to tell things apart when they all look the same?

Learning Goal
Students will become aware of the importance of their sense of smell when confronted with objects that look the same.

Guiding Document
Project 2061 Benchmark
- *People use their senses to find out about their surroundings and themselves. Different senses give different information. Sometimes a person can get different information about the same thing by moving closer to it or further away from it.*

Science
Life science
human body
senses

Integrated Processes
Observing
Comparing and contrasting
Predicting
Applying

Materials
For the class:
extracts (examples: peppermint, lemon, orange)
clay dough (see *Management 1)*
red food coloring
chart paper

For each student:
white paper, 8.5 x 11-inch

Background Information
Our sense of sight often influences what we think things are. To convince young students that things are often not what they seem, it is necessary to challenge them with learning situations in which there is an imbedded surprise. Presenting young learners with items that are similar in appearance and texture, but have a different scent, is one way to do this. In this activity, students see and feel clay dough. It is then infused with various scents to help learners realize the importance of their sense of smell.

Management
1. Prior to the activity, make several batches of clay dough according to the recipe. Allow time for it to cool and cure. One recipe will make five balls of clay dough the size of tennis balls. You will need three balls for each group of six students.

Clay Dough

3 c flour
1 ½ c salt
3 c water
2 Tbsp oil
3 tsp cream of tartar

- Cook over low heat, stirring constantly until mixture is the consistency of mashed potatoes, and it begins to lump.
- Remove from heat and knead until a dough-like consistency is reached.
- Divide dough into balls the size of tennis balls.

2. Most dough will be used in its natural color; however, reserve a tennis-ball sized portion for each group. Color the reserved dough with red food coloring and scent it using lemon extract.
3. It is suggested that one group of six students work at this station while other students are engaged at other stations working on other activities.

4. Prepare a chart with the title *Making Sense of What You Smell.* Place several markers in the area for the recording of students' guesses.

Procedure

1. Introduce the uncolored, unscented clay dough to the group of students at the station. Have them look at and touch each of three balls to decide that they look and feel the same.
2. Make a well in each of the three balls of clay dough. Put 3-5 drops of one of the extracts into one ball. Knead the ball until the scent is spread throughout.
3. Follow the same procedure for the other two balls using the other two extracts.
4. Give each student a piece of paper. Tell them to fold it in half and then fold it in half again so they have 4 spaces. Have them number the spaces 1-4.

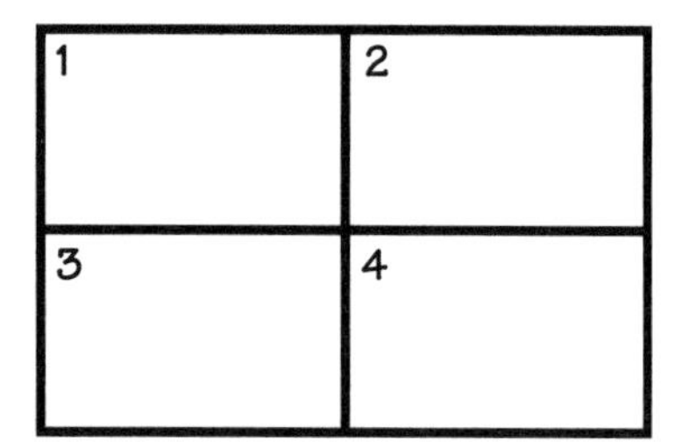

5. With one ball of scented clay dough, pinch off enough to give each student in the group a walnut-sized piece. Tell students to place this first clay dough ball in area numbered 1 on their paper. Do the same with the second ball of dough, having the students place their pieces in the area numbered 2, etc. Tell students not to mix the dough balls together.
6. Have the students smell the clay dough in area 1. Record their guesses on the chart paper as to what they think they are smelling. Continue with the clay dough in areas 2 and 3.
7. Show the students the clay dough you have colored and scented prior to class time. Ask them to guess what the scent will be by simply looking at it. Record the guesses on the *Making Sense of What You Smell* chart.
8. Give the students walnut-sized pieces of the scented red clay dough to smell. Have them use area 4 on their sorting mats for this clay dough ball.
9. Ask students if they want to change their guesses and allow them to do so, if they wish.
10. After discussion, allow students to mold objects of their choice from the clay dough. Set them aside to dry.

Connecting Learning

Using the unscented, non-colored clay dough balls:

1. Look at the clay dough. Do the three balls look the same?
2. Smell the clay dough. What does it smell like? Do all three balls of clay dough smell the same?

Using the scented, non-colored clay dough balls (one at a time):

1. Smell the clay dough. Do the three balls smell the same?
2. What scent do you think you are smelling? What does it remind you of? (Continue until the students have examined all three clay dough balls.)
3. Do all the clay dough balls look the same or different? How can you tell the difference between the three balls of clay dough? [by smelling them]
4. What sense did you use to name the scent of the clay dough balls? [sense of smell]

Using the colored, scented clay dough ball;

1. After smelling the red clay dough ball, did you change your prediction or leave it the same?
2. Did the color help you make your prediction? Explain.
3. What happened when you used your sense of smell?
4. Did you guess the correct scent when you used just your sense of sight? ...your sense of smell? ...your senses of sight and smell?
5. If you had some clay dough that was orange and smelled like bananas, would you be able to name the scent? What does your sight do to your sense of smell?
6. Does your sense of smell or your sense of sight tell you what the scent of an object is?
7. If an object seems to be the wrong color, does it confuse our sense of smell?
8. What are you wondering now?

Making Scents From Scratch

(A Directed Art Activity)

Topic
"Scratch and sniff" art project

Materials
Flavored gelatin and/or flavored crystal drink mix
White glue
Student page
Plastic spoons
Plastic cups, 9 oz
Various flavors of extracts, optional

Management
1. Set up a station or center for students to work independently or in small groups.
2. Put the necessary flavors of the dry gelatin in 9-ounce plastic cups with a plastic spoon in each one. You will need orange, lemon, lime, grape, pineapple, strawberry, raspberry, and cherry.
3. Set out several bottles of white glue.
4. Either distribute student page (one per student) prior to the activity, or provide an ample supply of student pages at the center for the students to use.

Procedure
1. Direct students to cover an area on their student art page with a thin layer of glue. Have them select the appropriate flavor of gelatin or crystal drink mix that depicts the fruit covered. For example: cover the strawberry picture with glue and then sprinkle strawberry gelatin all over the surface of the glue. Allow to dry.
2. Once the glue has dried, a drop of an extract can be added to enhance the scent. Again, allow to dry.
3. Continue this process until all the pictures of fruit have been covered with the gelatin and glue.
4. Once the entire project is dried, direct the students to gently scratch the surface of one of the fruits and to then smell the scent.

Hearing

It would seem that our hearing is reasonably sensitive. After all, we can adjust quickly to a great range of amplifications, from the sound of industrial noise to a friend's whisper a few seconds later. We can also hear a wide range of frequencies, from the low notes of a pipe organ to the high tones of a flute or piccolo. There are, however, many sounds that we humans cannot hear that other animals can. Bats have a much larger frequency range than humans. They use high frequency pitches to echolocate. Dogs can be summoned by blowing high frequency whistles that humans cannot hear. Animals such as donkeys and rabbits have large outer ears that they can move to pinpoint the source of a sound. We can do the same, although less precisely. If we hear a noise, we can turn our heads to pick up the strongest sound waves.

Hearing is important to everyday life. A large part of learning involves hearing. We learn to talk with all the proper inflections by listening to others, but people born with severely limited hearing have difficulty in learning to talk. Not only can they not hear the speech of others, but they cannot hear their own voices.

Human ears come in a remarkable variety of shapes and sizes. The outer flaps may not appear very complicated, but they play a very important role in how we interact with our world.

The outer ear flaps, called auricles, catch sound waves and funnel them into the auditory canal where a chain of events occur. The sound waves travel through the auditory canal where they strike the eardrum (tympanic membrane) and cause it to vibrate. The vibrations are passed on to three tiny bones in the middle ear called the hammer (malleus), the anvil (incus), and the stirrup (stapes). These bones transmit the vibrations to a membrane of the oval window that covers the opening of the inner ear.

The receptor cells for hearing are located within the inner ear. They are found in the cochlea, a fluid-filled tube that contains hair-like receptors that respond to vibrations. The receptors send nerve signals to the brain where they are perceived as sound.

The hearing centers of the brain are located on both sides, just above the level of the ear. The brain begins building sound memories before birth. A baby in the womb can hear the sounds of its mother's heart beating and louder noises from outside. Gradually, sounds are linked with events; we connect the honking of a car's horn with danger, the creak of a door with someone entering the room, etc. Sounds are sorted out by the brain and compared with those in the memory. The brain then decides if the sound is important enough to act upon.

A Dryer Full of Tennis Shoes

When sounds are all around you,
It's hard to pick and choose.
You have to write a poem on sound—
What noises would you use?

The television's playing,
Someone's knocking at the door.
The dog is right beside you
Sleeping loudly on the floor.

A dryer full of tennis shoes
Is running in the hall;
The buzzer buzzes, cycle's done
You hear the gym shoes fall.

Your baby brother's crying,
Your pet parrot's in a rage,
He's squawking at the gardener
And pacing 'cross his cage.

Your mother yells, she'll sell him
If he doesn't quiet down.
You say your brother or the bird?
Imagining her frown!

A car horn says you're running late,
A clock's Westminster chime,
You're due at school in minutes
And your poem better rhyme!

Brenda Dahl

"chirp"
"chirp"
"chirp"
What Can I Hear?
Words by Suzy Gazlay
Tune: Billy Boy
Oh, what can I hear with my
ears, with my ears? Oh, what can I
hear when I listen? Child-ren laugh-ing, birds in
trees, pass-ing cars, and buzz-ing bees: These are
sounds that I hear when I listen.

Oh, what can I hear with my ears, with my ears?
Oh, what can I hear when I listen?
Voices on the radio,
Someone walking very slow;
These are sounds that I hear when I listen.

Oh, what can I hear with my ears, with my ears?
Oh, what can I hear when I listen?
Airplanes flying overhead,
Something nice that someone said;
These are sounds that I hear when I listen.

Oh, what can I hear with my ears, with my ears?
Oh, what can I hear when I listen?
Raindrops falling on my house
And the squeaking of a mouse;
These are sounds that I hear when I listen.

Topic
Sense of hearing

Key Question
What senses do we use to discover the contents of a sealed box?

Learning Goal
Students will use their sense of hearing to predict the contents of a set of sealed boxes.

Guiding Documents
Project 2061 Benchmark
- *People use their senses to find out about their surroundings and themselves. Different senses give different information. Sometimes a person can get different information about the same thing by moving closer to it or further away from it.*

*NCTM Standard 2000**
- *Represent data using concrete objects, pictures, and graphs*

Math
Graphing

Science
Life science
human body
senses

Integrated Processes
Observing
Comparing and contrasting
Predicting
Collecting and recording data
Interpreting data

Materials
Part One
For the class:
5 or more boxes with lids (examples: children's size shoe boxes, cheese boxes)
5 pairs of small objects (see *Management 1)*
class charts
patterned wrapping paper or adhesive plastic, optional

For each student:
glue
student page
prediction markers

Part Two
For each group:
small opaque container (see *Management 4)*
2 paper clips
2 marbles
2 small wooden blocks

Background Information
Hearing is the sense that is second only to sight in the degree of development in humans. Our senses can be explored by isolating them and focusing on how each works. We can help students to fine-tune the use of their senses to aid in the discovery and learning of the world around them. Through this activity the students focus on hearing to discover information about objects that cannot be seen and are not touched.

Management
1. Collect five pairs of the small objects (toy cars, pennies, balls, jacks, Unifix cubes).The duplicate object provides a point of reference as a visual clue for matching.
2. Beforehand, place a different object in each of the five boxes. Seal the boxes and number them, keeping record as to which object is in each box. To look more like gift boxes, you may wish to cover the boxes with wrapping paper or adhesive wrap.
3. This investigation can be used with the whole class or with small groups.
4. Collect enough small opaque containers with lids (such as yogurt or cottage cheese containers) so that each group of students can have one.
5. Cut prediction markers (the pictures of the ball, the car, a jack, a penny, and the Unifix cube) so each student, or group, will have the picture set of the five objects.

Procedure

Part One

1. Read one of the suggested books listed in *Curriculum Correlation.* Discuss how difficult it is to wait to open gifts. How could you determine what your gifts are without opening them?
2. Show one of the boxes.
3. Guide students to decide which sense they could use to discover what is in the box. Review which senses they would be using and discuss the body part(s) associated with those senses.
4. Give each student a turn to handle the box and listen to the sounds the contents make as the box is moved back and forth.
5. Ask the students what, if anything, they can tell about the contents from just moving the box and listening.
6. Introduce the set of five objects to be used as a point of reference. These five objects **must** be identical to the five you previously wrapped in boxes.
7. Ask the students to observe the objects. Guide them to use the process of elimination in selecting their predictions of what could be in the box. By a show of hands, have students indicate what they think is in the first box.
8. Discuss the number of students predicting each object for the box shown. Direct students to place the picture of the predicted object in *Box 1* on the activity page.
9. Have a student open the box. Ask the students to check their predictions with the actual. If they were not correct in their predictions, instruct them to replace the markers with the correct markers. Direct all students to glue down the correct marker in *Box 1.*
10. Repeat the activity with other objects. Be sure to allow each student to handle the box, move it around and to listen to the sound made by the object in the box. Each time the options will get fewer. For *Box 5,* ask the students why they do not have to listen to the sounds the object makes.
11. Prepare a chart for the final actual results. You may want to use the prediction markers and boxes from the student pages for the legends of the chart. After the box is opened, record what was actually in the box. Use pictures or Xs.

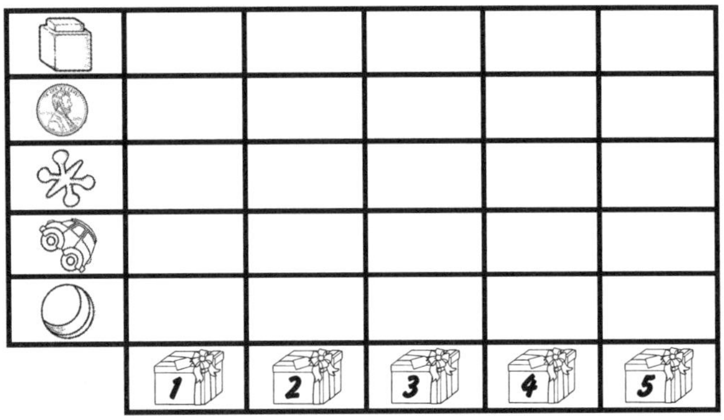

	1	2	3	4	5

Part Two

1. Divide students into groups. Give each group the materials.
2. Ask one student to secretly place one of the three objects—paper clip, marble, or wooden block in the box. (Again, one set of the three objects will be displayed so the students have a reference set). Have the student gently move the box back and forth, allowing the other students to listen, but not to touch.
3. Ask the other students to predict what they think is in the box.
4. Once all students in the group have had a chance to express their predictions, direct the student who packed the box to open it and reveal the actual object.
5. You may want to use the student recording pages to either record their predictions or the actual objects.

Connecting Learning

Part One

1. What clues were the most helpful in discovering the contents of the boxes? [having objects as a frame of reference]
2. What sense did we use most to discover the contents?
3. How could we make the discovery easier? ...harder?
4. Why do you think the _______ was easier to determine than the _______?
5. Why do you think the _______ was the hardest to determine?
6. Look at the shapes of the objects we used. What makes them easier (harder) to determine?
7. What other objects could we have used that would have been easy to predict by just listening to them in our boxes. [bells]
8. What other objects could we have used that would have been hard to predict by listening to them?

Part Two

1. Was it harder to predict what was in these containers than it was in the boxes with the cars, pennies, balls, jacks, and Unifix cubes? Why or why not?
2. Was it easier when you got to handle the containers and feel the objects move? Explain.
3. What sense did you use to determine what was in the containers?
4. What are you wondering now?

Extensions

1. Have students make their own "secret boxes" to share with the class.
2. Discuss how different this activity would be if you had difficulty hearing. Apply this to everyday life for the hearing impaired. This could lead to learning about similarities and differences of people.
3. Ask one student to hide behind a closed door. Instruct another student to listen to the hidden student's voice and determine who is behind the door.

Curriculum Correlation

Carle, Eric. *The Secret Birthday Message.* Harper Collins. New York. 1986.

Child, Lauren. *You Won't Like This Present as Much as I Do!* Grosset & Dunlap. New York. 2009.

Carter, David A. *How Many Bugs in a Box?* Little Simon. New York. 2006.

Carter, David A. *More Bugs in Boxes.* Little Simon. New York. 1990.

Zolotow, Charlotte. *Mr. Rabbit and the Lovely Present.* Live Oak Media. Pine Plains, NY. 1991.

Home Links

1. Have the students play a guessing game with their parents. With one family member behind a door, ask another person to listen to the voice to try to determine who is hiding. The same game can be played using objects that make noise.
2. Create "secret boxes" as a family project to share at home or take to school.

Secret
Sounds
1
2
3
4
5

Secret Sounds Prediction Markers

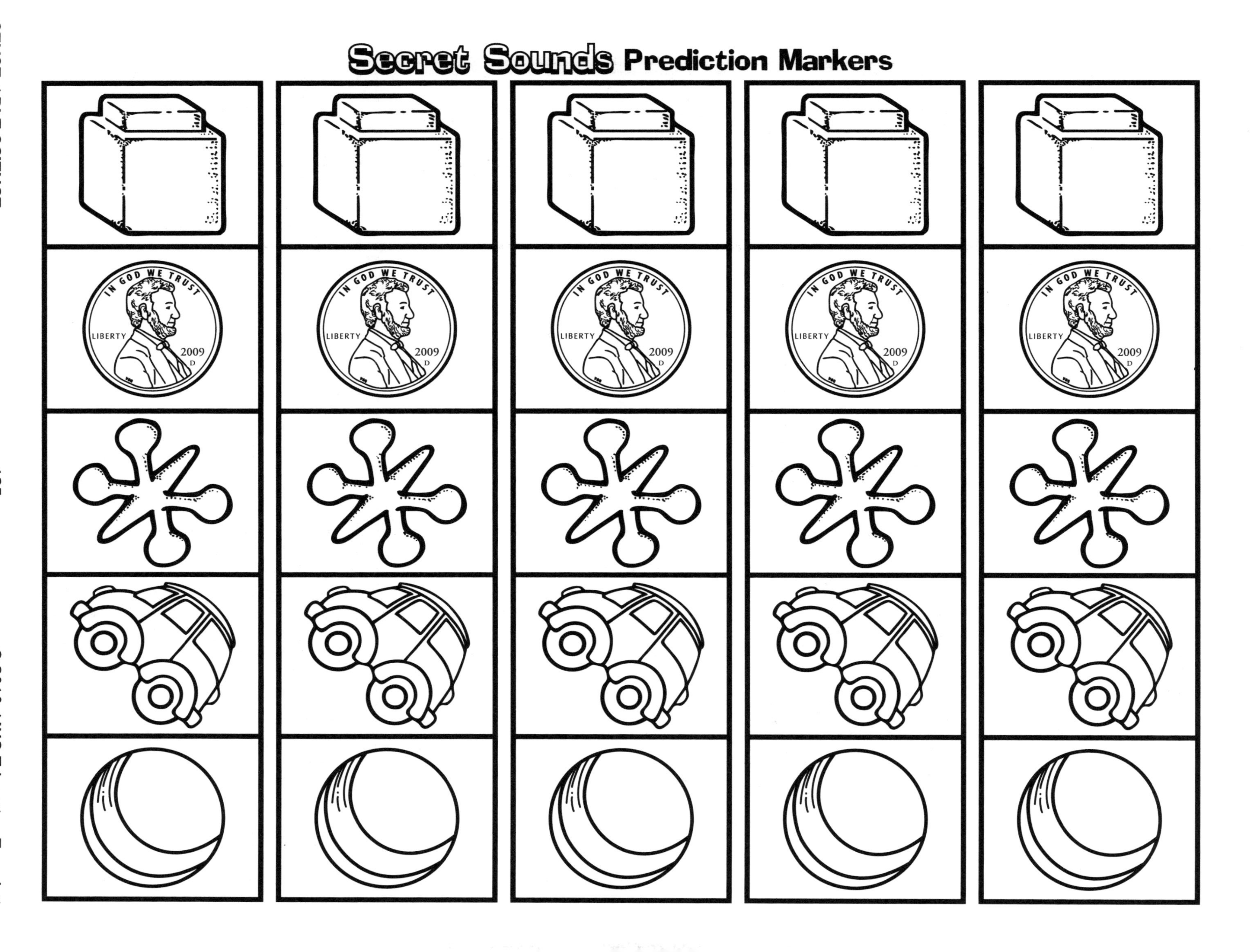

Secret Sounds

WALK, STOP, and LISTEN

Topic
Sense of hearing

Key Question
What can our ears tell us about our school?

Learning Goal
Students will become aware of information they gather through their sense of hearing.

Guiding Document
Project 2061 Benchmark
- *People use their senses to find out about their surroundings and themselves. Different senses give different information. Sometimes a person can get different information about the same thing by moving closer to it or further away from it.*

Science
Life science
  human body
    senses

Integrated Processes
Observing
Comparing and contrasting
Classifying
Collecting and recording data
Interpreting data

Materials
For the class:
  markers
  class chart, enlarged

For each student:
  personal blindfold (see *Management 5)*
  paper
  crayons

For each cross-age tutor:
  recording sheet (see *Management 3)*
  pencil
  personal clipboard (see *Management 6)*

Background Information
The brain begins building sound memories before birth. A baby in the womb can hear the sounds of its mother's heart beating and louder noises from outside. Gradually, sounds are linked with events; we connect the honking of a car's horn with danger, the creak of a door with someone entering the room, etc. Sounds are sorted out by the brain and compared with those in the memory. The brain then decides if the sound is important enough to act upon.

Management
1. Cross-age tutors are necessary for this activity. The tutors will serve as sighted guides and recorders for the younger students, who will be blindfolded. The older students will lead the younger students to four different areas around the school. The younger students will walk one step behind their sighted guides, holding their guides' arms just above the elbows.

2. Select four different areas around the school where students can go to listen. Suggestions include the office, cafeteria, all-purpose room, and library. Inform the cross-age tutors of the four locations that have been selected.
3. Prepare recording sheets by folding paper into fourths and numbering each section. Instruct the sighted guides that each section is a designated area that will be visited. Assign a number, one through four, to each area. The guides will record the younger students' responses from each area in the appropriately numbered section of the paper.
4. Enlarge the class chart showing one section for each of the areas that have been selected for visiting.
5. If personal blindfolds have not been previously made, make them prior to doing this activity.
6. The sighted guides will need a hard surface (book or clipboard) on which to write during the blindfolded tour.

Procedure

1. Practice listening with students in the classroom. Discuss the many sounds they hear.
2. Rehearse the sighted-guide technique with the students without using the blindfolds.
3. Distribute recording sheets, pencils, and clipboards to the sighted guides.
4. Explain that the sighted guides will blindfold their younger student partners and escort them to four specified locations. At each location, they will stop and sit still for two minutes to listen. The younger students will tell their guides what they hear and where they think they are. (The guides will not confirm or deny the location.) The guides will record the responses for each location in the correct numbered section on the recording sheets.
5. Send the student pairs out to the four locations where they will be listening. Be sure the guides know that not all pairs of students need to visit the locations in the same order.
6. When student pairs return to the classroom, have the younger students remove their blindfolds. Ask the guides to share the things that were heard by the younger students in each of the four areas and their guesses as to the locations. Make lists of the sounds as they are being shared.
7. Have the younger students draw pictures of whatever made the sound that they liked best.
8. Post the pictures on the class chart in the appropriate area (1, 2, 3, or 4) where the sound was heard. Identify each of the four locations.

Connecting Learning

1. What did you hear?
2. What can you tell about stop number one? ...two? ...three? ...four?
3. How were the sounds at stop one different from the sounds at stop two?
4. How were the sounds the same?
5. Did you need your eyes to tell where you were? Explain.
6. By looking at our chart, can we name the places we went?
7. At which area did our class find their favorite sounds? ...the least favorite sounds? Why do you think this was so?
8. What are you wondering now?

Extensions

1. Take the students outside, blindfold them, and have them sit on the playground for a few minutes before recess begins. Discuss the sounds. Have them continue to sit while the bell rings and other classes come outside for recess. Discuss the differences heard before, during, and after recess.
2. Have the students go into the cafeteria during lunch. Seat them at a table, put on blindfolds, and listen to the sounds of the cafeteria as the different classes come and go. After lunch, discuss all the sounds the students can recall hearing.
3. Record an entire listening walk. After returning to the classroom, listen to the recording and discuss what made the sounds.

Curriculum Correlation

Language Arts

Write about the sounds. Use sentences such as: My favorite sound was ______. I did not like the ______ sound.

Social Studies

Discuss other sounds heard in the home and around the community.

Home Link

Give students two 3" x 5" cards or pieces of paper to take home. Have them draw their favorite home sounds and bring the pictures back to class to make a class chart.

Patterns for Student Blindfolds

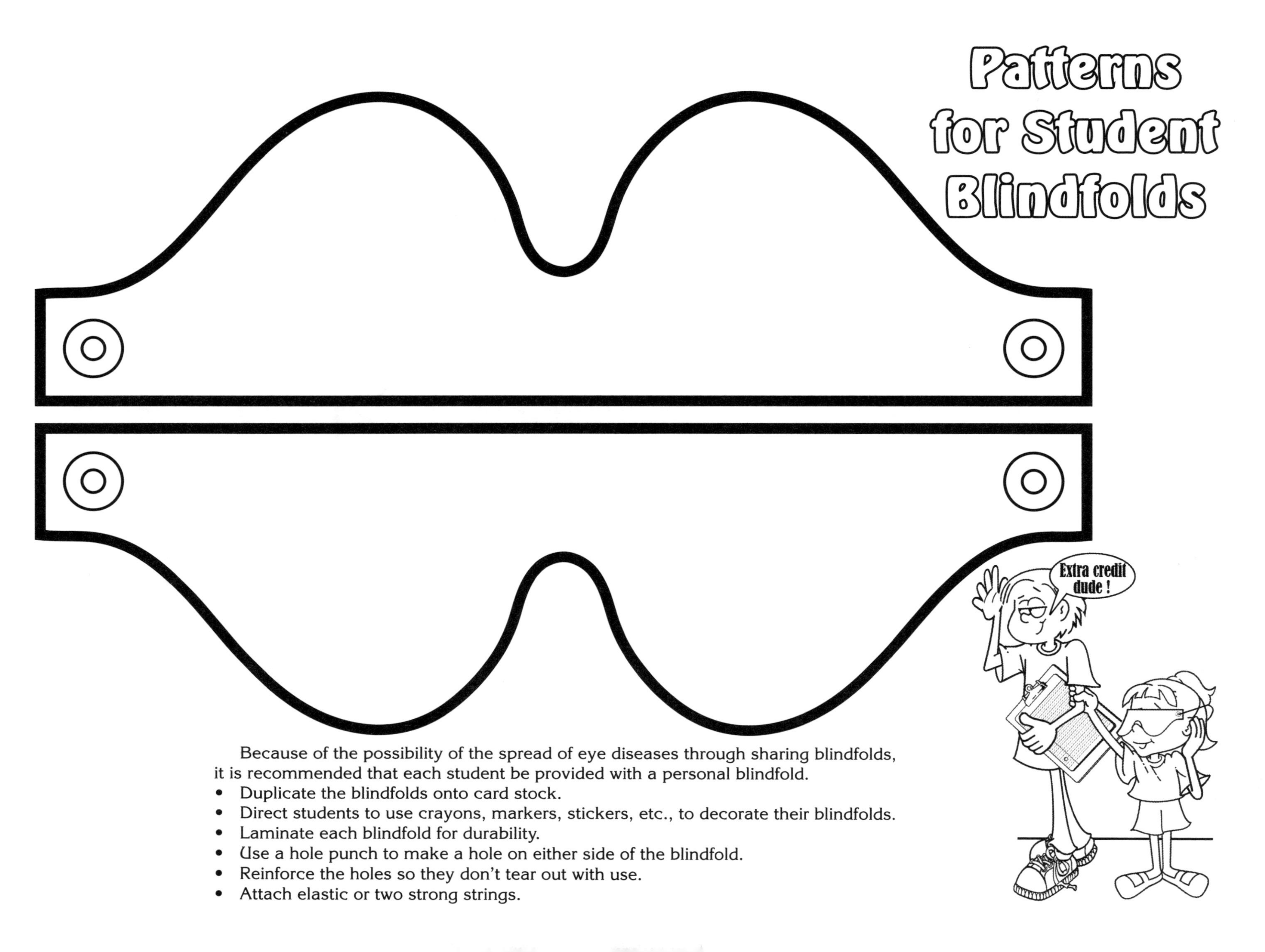

Because of the possibility of the spread of eye diseases through sharing blindfolds, it is recommended that each student be provided with a personal blindfold.

- Duplicate the blindfolds onto card stock.
- Direct students to use crayons, markers, stickers, etc., to decorate their blindfolds.
- Laminate each blindfold for durability.
- Use a hole punch to make a hole on either side of the blindfold.
- Reinforce the holes so they don't tear out with use.
- Attach elastic or two strong strings.

Our Listening Chart

1	2	3	4

Designer Ears

Topic
Sense of hearing

Key Question
Does the size and shape of ears make a difference in how well we hear?

Learning Goals
Students will:
- use cups to make designer ears, and
- determine whether the cups help them to hear better when held in a variety of positions.

Guiding Documents
Project 2061 Benchmark
- *People use their senses to find out about their surroundings and themselves. Different senses give different information. Sometimes a person can get different information about the same thing by moving closer to it or further away from it.*

NRC Standards
- *Each plant or animal has different structures that serve different functions in growth, survival, and reproduction. For example, humans have distinct body structures for walking, holding, seeing, and talking.*
- *The behavior of individual organisms is influenced by internal cues (such as hunger) and by external cues (such as a change in the environment). Humans and other organisms have senses that help them detect internal and external cues.*

Science
Life science
 human body
  senses

Integrated Processes
Observing
Comparing and contrasting
Inferring
Applying

Materials
Kitchen timer that ticks
Paper cups, 9 oz
Scissors
Animal pictures

Background Information
Sound travels through the air in waves that spread out from the source of the sound. When a sound is produced, some of the sound waves will reach your ears, but most of them are spread out into the surrounding area. When you cup your ear with your hand, you are actually channeling more of the sound waves into your outer ear.

Many animals have ears that are better designed to pick up sounds than are human ears. These animals tend to have more cup-shaped and/or proportionally larger ears than we humans have. Many also have the advantage of being able to turn their ears to help them locate the direction of sounds. Humans must turn their heads to do this.

Management
1. Prior to this activity, make ears for each student by cutting a 9-oz paper cup in half and removing the bottom.

2. Plan to display the page of animal pictures using a projection device.

Procedure
1. Show students the kitchen timer and ask what will happen if you turn the dial. [It will make a ticking noise.]
2. Stand at the front of the room and tell students that you want them all to turn their chairs so that they are facing the front. Turn the dial and allow students to hear the noise of the timer. Ask what sense and body part they are using. [hearing, ears]
3. Tell the students that you want them to continue facing forward and to pay attention to the sound of the timer. Walk to the back of the class with the timer. Ask students what they notice. [The timer sounded louder when you walked by me; it's harder to hear the timer when it's behind me; etc.]

4. Turn off the timer and show students the page of animal pictures. Ask if they think any of these animals would be able to hear the timer better than they could, and why. Explain that many animals have larger ears (relative to their size) than humans do and are able to move those ears to follow the direction of a sound.
5. Explain that students will be testing some "designer ears" to see if they help them better hear the kitchen timer.
6. Distribute two ears to each student. Allow them to have a few minutes of free exploration where they hold the cups to their ears in various ways and see how they change the sounds they hear.
7. Take the kitchen timer to the front of the room again and have all students turn to face you. Point out the picture of the elephant and ask students to hold their designer ears like the elephant's ears (straight out from the sides of their head, with the cupped parts pointing forward). Turn on the timer and have students listen for a few seconds, then remove their designer ears and compare the sound. Discuss whether the designer ears improved their ability to hear the sound.
8. Repeat this process with additional animals, and move around the room, allowing students to experiment with moving the designer ears (but not their heads) to hear things in different locations relative to their ears.
9. Close with a discussion about what students learned about our ears and how their size and shape affects how well we hear.

Connecting Learning

1. What body part do you use to hear? [ears]
2. What happened to the sound of the kitchen timer as it was moved around the room? [It was easiest to hear when it was close to and in front of us.]
3. Are you able to turn your ears to follow a sound? [No.] What do you have to do instead? [turn my head/body]
4. How are our ears different from the ears of animals? [Many animals have larger ears that they can move.]
5. What did you discover when you used your designer ears? Did they improve your ability to hear the kitchen timer?
6. Which designer ear position let you hear the timer best? Did it matter where the timer was?
7. Were there any ways that you held your designer ears that made your hearing worse? [Holding the cups over the top of the ears (like the pig's or dog's floppy ears) made it harder to hear.]
8. Why do you think the outside parts of our ears are shaped the way they are? [They help direct sounds into the inside of the ear.]
9. What are you wondering now?

Extensions

1. Allow students to try additional designs for their ears using a variety of materials. Have them compare the effectiveness of the different ear styles.
2. Study animals and their hearing. Research information about animals that have a keen sense of hearing compared to those that do not. What are the differences in the sizes and shapes of their ears?

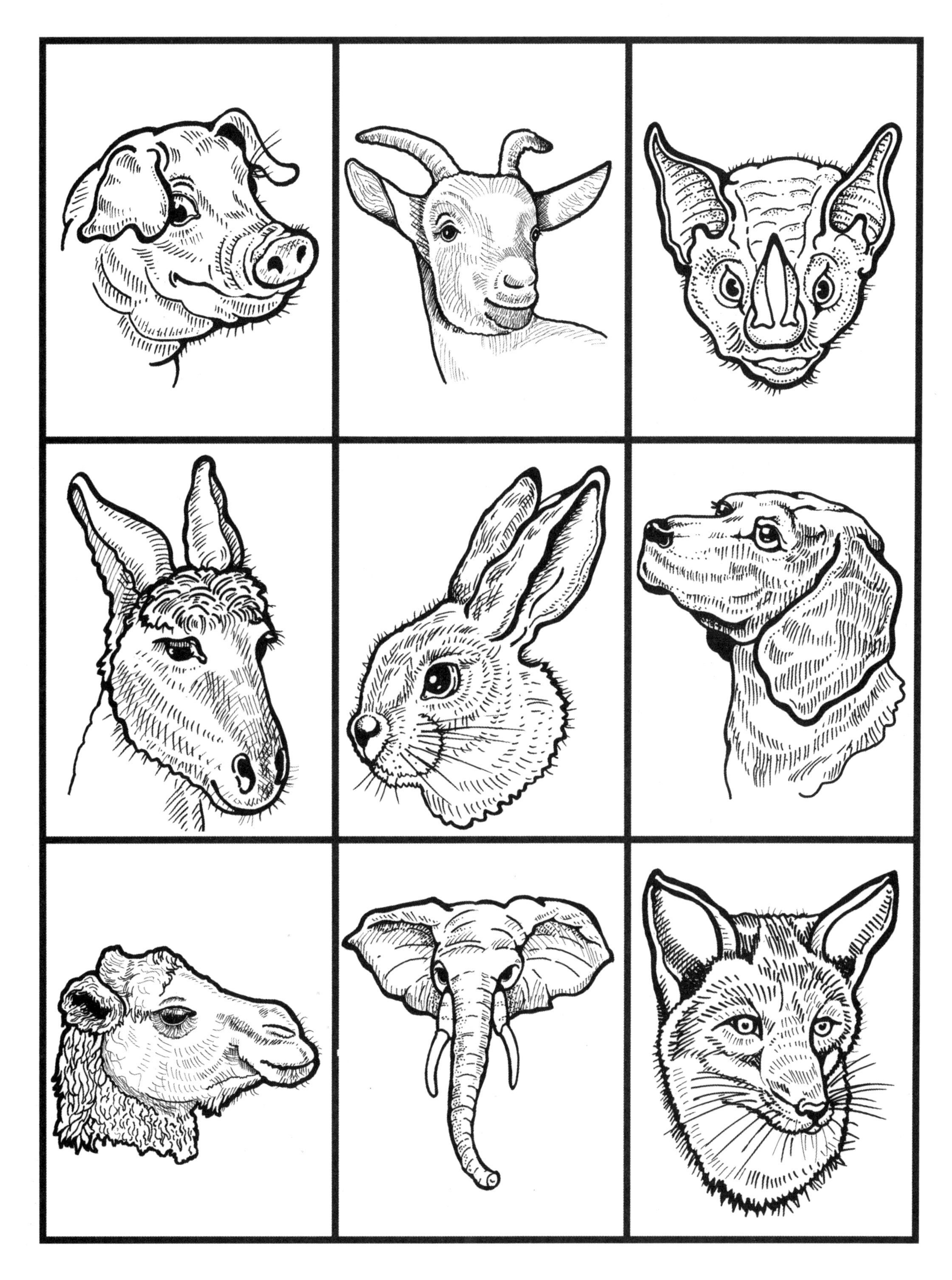

Topic
Sense of hearing

Key Questions
1. Do we really need both ears to hear?
2. Do we need both ears to locate where the sound is coming from?

Learning Goal
Students will discover that they can hear sounds using only one ear, but that it is easier to locate the source of sounds around them when they use both ears.

Guiding Documents
Project 2061 Benchmark
- *People use their senses to find out about their surroundings and themselves. Different senses give different information. Sometimes a person can get different information about the same thing by moving closer to it or further away from it.*

*NCTM Standard 2000**
- *Count with understanding and recognize "how many" in sets of objects*

Math
Counting
Tallying

Science
Life science
human body
senses

Integrated Processes
Observing
Comparing and contrasting
Collecting and recording data
Interpreting data
Inferring
Applying

Materials
For each group of four students:
2 pennies
transparent tape
2 crayons of different colors

For each student:
student blindfolds
rabbit tally piece

Background Information
When you cover one of your ears, you can still hear. Then why do we have two ears? This activity will demonstrate to the students that they can hear with only one ear, but the two ears working together help them to determine where the sounds they are listening to are coming from.

When you are listening with both ears, sounds reach the ear closest to them a split second before reaching the other ear. In addition, the sound in the closer ear is slightly louder, since the head interferes with the sound reaching the other ear. When one ear is covered, you cannot usually use these slight differences in sound speed and volume to locate the sound source.

Humans have binaural hearing; that is, we use two ears to hear. People with hearing in only one of their ears many have difficulty locating sounds. They might need to turn their heads in several directions before finding the source of the sound.

Management
1. It is not recommended that students share blindfolds because of the possible exposure to eye diseases. If personal blindfolds are not already made, use the pattern and make prior to doing this activity.
2. This activity begins with a demonstration of procedures. The class is then divided into groups of four. One student will be the *Direction Detector,* another student will be the evaluator to determine whether the guess is *Right On* or *Way Off,* a third student will be the recorder, and a fourth student will be the castanet player.

3. The clapping of penny castanets will be used as sound sources. Penny castanets are made by taking a small strip of tape (large enough to fit around the end of your finger) and making a circle with the sticky side on the outside. A penny is then stuck onto the tape. Each group will need two of these penny devices. One will be placed over the thumb and one over the forefinger of the same hand.

4. Copy the tally records, one per student. Students will record by tallying the *Direction Detectors* results. *Right On* will be tallied on the rabbit's right ear. *Way Off* will be tallied on the left ear.
5. Set standards for determining whether guesses are *Right On* or *Way Off.*

Procedure

1. To demonstrate the procedure for the class, have students sit in a large circle. Select a student, the *Direction Detector,* to sit in a chair that is placed in the middle of the circle.
2. Blindfold the *Direction Detector* making certain not to cover his/her ears. Caution students to be very, very quiet because you will be using penny castanets to make a sound. The *Direction Detector* will then try to point in the direction of the location of the sound. Tell the *Direction Detector* not to turn his or her head.
3. Place the castanets on your forefinger and thumb. Quietly walk to a location in the room and click the pennies twice. The *Direction Detector* is to point in the direction he/she believes you are standing. Continue by moving to different locations.
4. To demonstrate the tallying component of this activity, mark each *Right On* guess by the *Direction Detector* on the rabbit's right ear and each *Way Off* guess on the left ear.

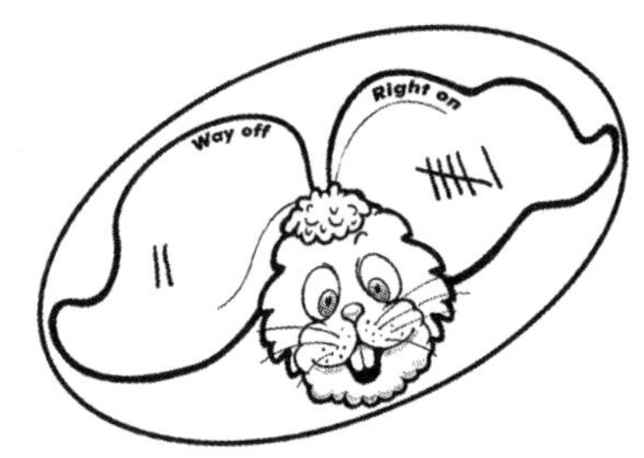

Review tallying by making four marks and showing the fifth mark as a diagonal across the existing four. Use one color for two-ear results. (The other color will be used for one-ear results.)

5. After several attempts, direct the *Direction Detector* to press one finger or the palm of his/her hand against one ear, closing off the ear canal and preventing sound waves from entering.
6. Repeat the above procedures, moving to different locations, clicking twice and recording the results with the second color.
7. Divide the class into groups of four. Allow students to rotate jobs until each student has an opportunity to be the *Detection Detector.*
8. Have each student experience the activity using both ears and then covering one. Record the results.
9. Tell students to tape their tally records on the wall. Have them count the number of *Right On* results and *Way Off* results. Record these results on the board.

Connecting Learning

1. What do our tally records tell us about this experience?
2. Were some people better than others at telling where the sound was coming from? Why might this be?
3. Were there more *Right On* responses or *Way Off* responses?
4. Was it harder to determine the location of the source of sound with only one ear? Why do you think it was that way?
5. What do you think people who can only hear out of one of their ears do to help them hear better? [They turn their heads toward the sounds to hear better.]
6. Why would it be important to know the source of a sound? [warnings of danger: sirens, barking dogs]
7. What are you wondering now?

Extension

Click the pennies directly in front of the blindfolded person. Do the same directly behind, and on top of his/her head. Is the person confused? When the sound comes from a spot midway between the ears, it reaches both ears at the same time, making it hard to figure out where the sound is coming from.

* Reprinted with permission from *Principles and Standards for School Mathematics*, 2000 by the National Council of Teachers of Mathematics. All rights reserved.

Patterns for Student Blindfolds

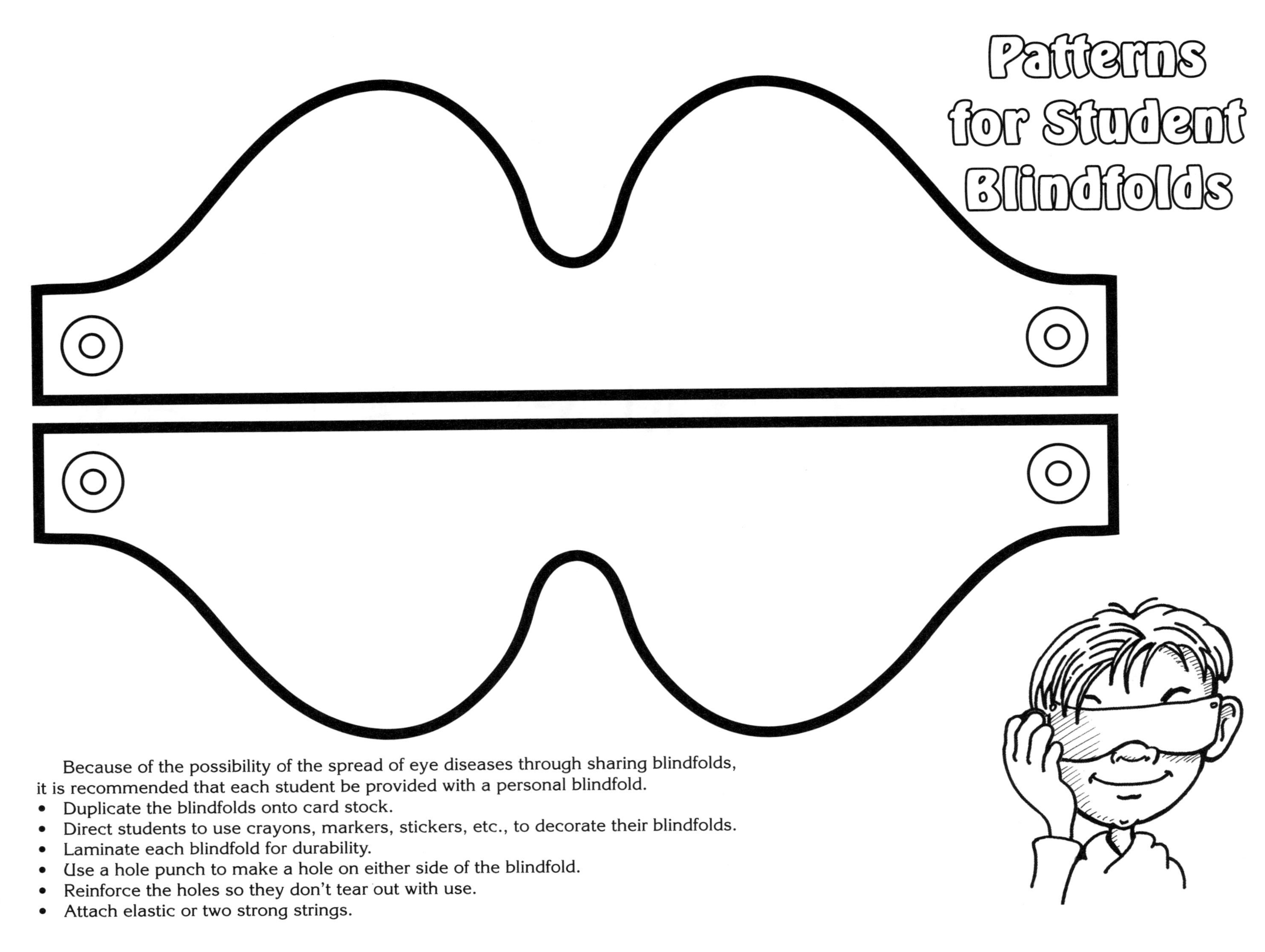

Because of the possibility of the spread of eye diseases through sharing blindfolds, it is recommended that each student be provided with a personal blindfold.

- Duplicate the blindfolds onto card stock.
- Direct students to use crayons, markers, stickers, etc., to decorate their blindfolds.
- Laminate each blindfold for durability.
- Use a hole punch to make a hole on either side of the blindfold.
- Reinforce the holes so they don't tear out with use.
- Attach elastic or two strong strings.

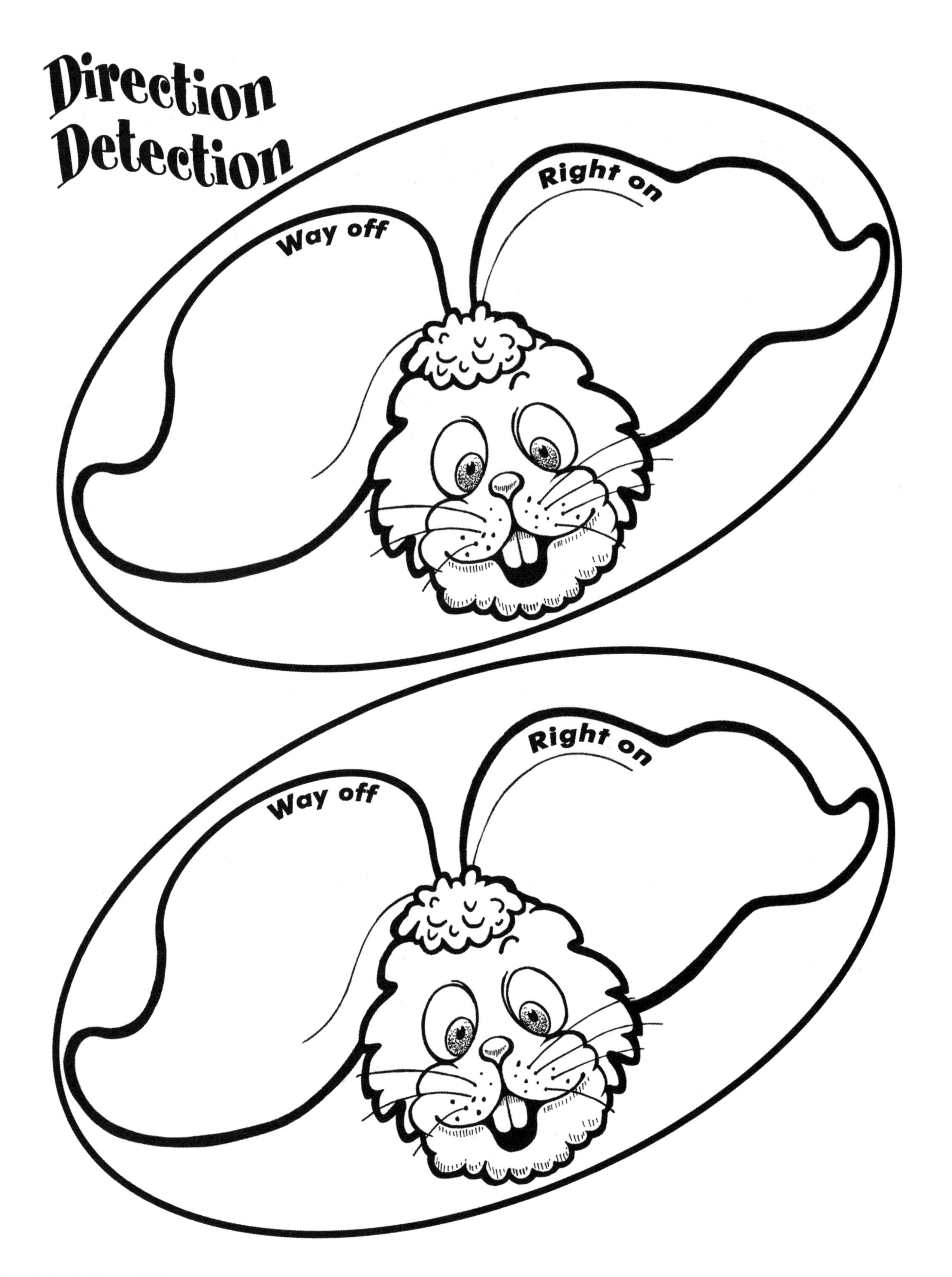
Direction Detection
Way off
Right on
Way off
Right on

Topic
Human body/senses

Key Question
How can we determine the sense/body part used to experience each object pictured?

Learning Goal
Students will match picture cards with sense cards.

Guiding Documents
Project 2061 Benchmarks
- *The human body has parts that help it seek, find, and take in food when it feels hunger—eyes and noses for detecting food, legs to get to it, arms to carry it away, and a mouth to eat it.*
- *Senses can warn individuals about danger; muscles help them to fight, hide, or get out of danger.*

NRC Standards
- *Each plant or animal has different structures that serve different functions in growth, survival, and reproduction. For example, humans have distinct body structures for walking, holding, seeing, and talking.*
- *The behavior of individual organisms is influenced by internal cues (such as hunger) and by external cues (such as a change in the environment). Humans and other organisms have senses that help them detect internal and external cues.*

Science
Life science
  human body
    senses
    body parts

Integrated Processes
Observing
Comparing and contrasting
Relating
Communicating
Interpreting data

Materials
Sense cards
Picture cards

Management
1. Copy and cut out one set of sense cards per group.
2. Copy and cut out the picture cards provided for each group. Additional pictures can be cut from magazines, if desired.

Procedure
1. Explain to the students that they will be working together to sort some picture cards and match them to the sense cards.
2. Show the students how to display the sense cards so that everyone in the group has a good view. Select a picture card and show students where you would place it. Be sure to give your reason for your choice. For example, a hair dryer goes with the ears because it makes a lot of noise. It could also go with hands and touching because it is held when used.
3. Divide students into groups and distribute the sets of cards. Have groups work together to put all of the picture cards with the sense cards. Encourage groups to discuss the multiple possibilities and to come to a consensus about where to put each picture.
4. When students have finished sorting all of the pictures, discuss how they decided which body part, or sense, the pictures went with. Encourage them to think about the multiple ways that each picture could be classified.
5. Collect the sense cards and the picture cards. If desired, place them in a center or station for independent practice.

Connecting Learning
1. How did you decide where to put each of the pictures?
2. Why could some of your picture cards have gone with different body parts/senses?
3. When a picture could go into more than one pile, how did your group decide what to do?
4. Can you think of something else that could go in the tasting pile? ...the hearing pile?
5. Do you think you use one of your senses more than another? Why or why not?
6. What are you wondering now?

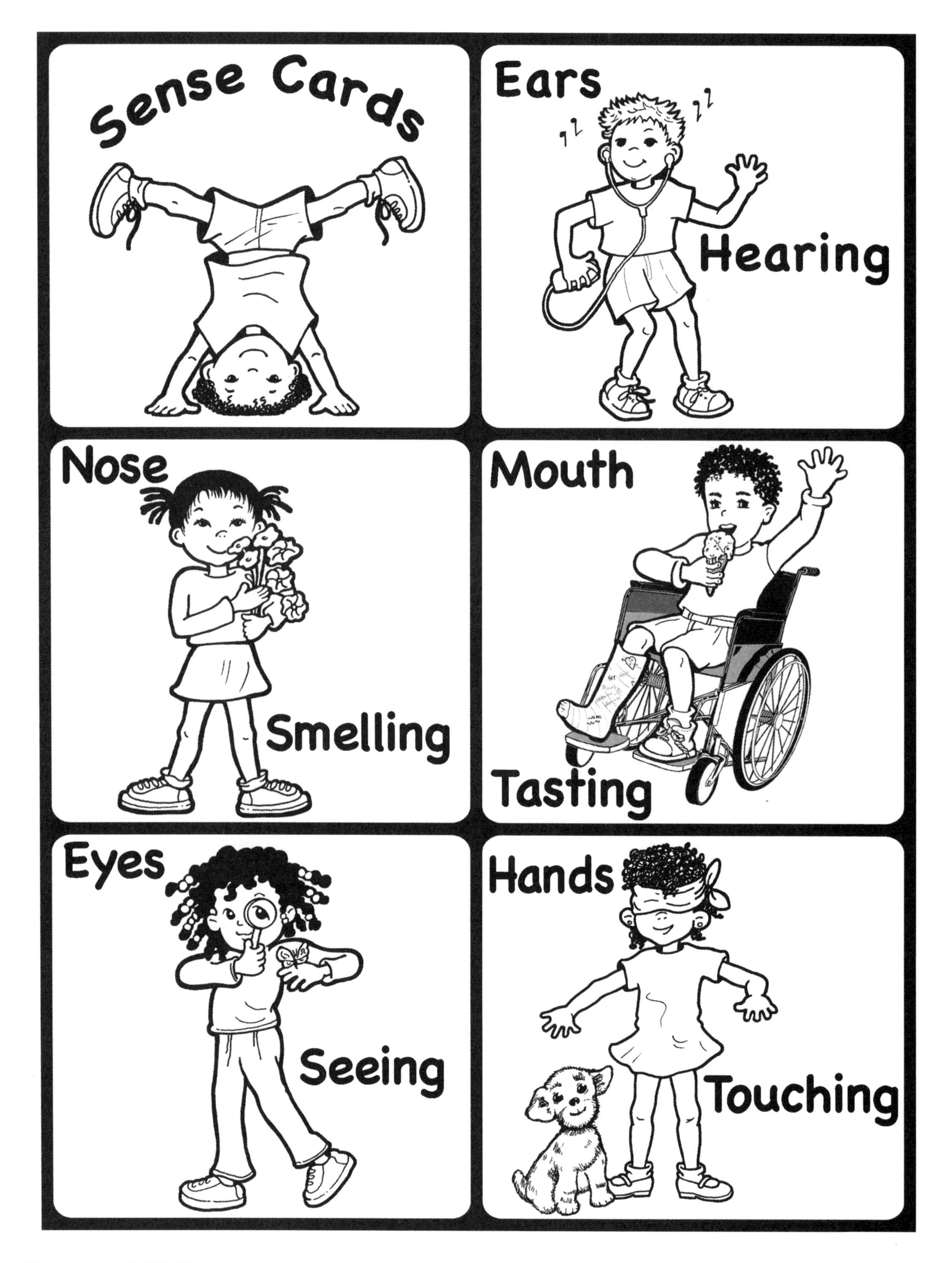
Sense Cards
Ears
Hearing
Nose
Smelling
Mouth
Tasting
Eyes
Seeing
Hands
Touching

0
5
10
15
20
25
30
35
40
45
50
55

On
Soda
Popcorn

Making Sense of Our Senses

Topic
Senses

Key Question
What body parts do we use to smell, hear, see, taste, and touch?

Learning Goal
Students will recognize the five senses and match them with the related body parts.

Guiding Documents
Project 2061 Benchmark
- *People use their senses to find out about their surroundings and themselves. Different senses give different information. Sometimes a person can get different information about the same thing by moving closer to it or further away from it*

NRC Standard
- *Each plant or animal has different structures that serve different functions in growth, survival, and reproduction. For example, humans have distinct body structures for walking, holding, seeing, and talking.*

Science
Life science
  human body
    senses

Integrated Processes
Observing
Predicting
Collecting and recording data
Interpreting data

Materials
Senses cards
Student page
8½" x 11" piece of card stock, one per student
Paper fasteners, 10 per student
Yarn
Hole punch
Glue sticks
Scissors

Background Information
Our senses help us gain information about our surroundings. We take the information in through our sensory organs. The sensory organs then send information to our brain. Visual data is collected by the eyes. We see colors and shapes; we perceive depth; we register movement. Our ears collect sounds, allowing us to hear noises made in the world around us. When we breathe through our noses, we register any scents or odors in the air. Taste buds on our tongues collect information about the flavors of the foods we eat and things we drink. Nerve endings in our skin respond to pressure, allowing us to feel things. Although the skin on most parts of our bodies is sensitive to touch, the hands are the body part generally associated with this sense.

It is important for young children to first identify the sense and sense organ that brings the information to their brain. They can then fine-tune the use of their senses to aid in more discovery and learning of the world around them. This activity begins with a discussion about the five senses and the related body parts and ends with them making a game to use as a review.

Management
1. Prior to teaching this lesson, prepare a game card for each student by punching five evenly-spaced holes down the right side of 8½" x 11" pieces of card stock as illustrated.

2. Students may need assistance in placing the paper fasteners into the game cards.

3. Copy the body parts page on card stock. Make enough copies for each student to have one set of body parts. The senses labels can be copied on plain paper.
4. Each child will need five 35-centimeter pieces of yarn.

Procedure

Part One

1. Lead the class in a discussion about the five senses and related body parts. Question the class to see what their prior knowledge on the topic is.
2. Give each student a set of body part cards. Ask them to cut the cards out and place them on the table in front of them.
3. Explain to the students that you will be asking them about their five senses and that they should hold up the body part that is related to the sense mentioned.
4. Ask the following questions:
   - When I taste a treat, I use my____
   - When I smell a rose, I use my _____
   - When I hear the bell, I use my _____
   - When I feel rough sand paper, I use my ____
   - When I see a rainbow, I use my ____
   - I smell with my ____
   - I hear with my ______
   - I see with my _______
   - I taste with my ______
   - I touch with my _____
5. End with a discussion about how they knew which body part was used for which sense.

Part Two

1. Review the five senses and related body parts.
2. Explain to the students that they will be making a sense game that they can use to review their five senses. Tell the class that the game will have the body parts we use to taste, smell, hear, see, and touch on one side of the game board and the matching words on the opposite side of the game board.
3. Distribute a set of senses labels, a pre-punched sheet of 8½" x 11" card stock, a glue stick, scissors, and paper fasteners to each student.
4. Instruct the students to cut out the labels and glue them beside the pre-punched holes in any order they wish.
5. Have students use the body part cards from *Part One* and glue them on the opposite side of the card stock. Remind students to mix the body parts up so that they are not always directly across from the correct sense word.
6. Assist students in placing the paper fasteners in the holes beside the words.
7. Give each child five 35-centimeter pieces of yarn and five pieces of clear tape. Help them tape the ends of the yarn to the right side of each part.

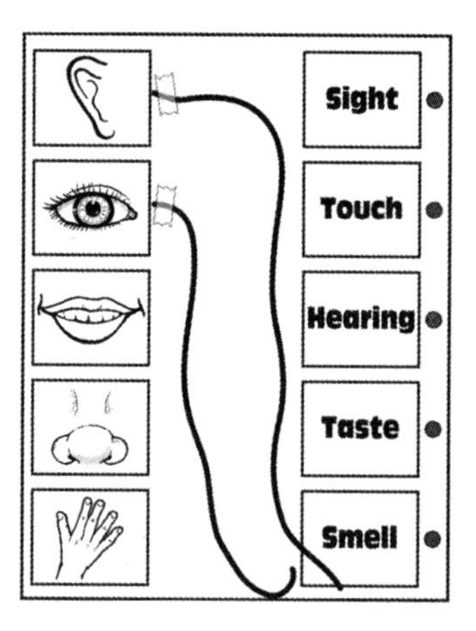

8. Using one of the student's game boards, demonstrate how to attach each body part to the related sense by wrapping the yarn around the paper fastener.
9. Allow the children to exchange game boards and test their skills.
10. End with a discussion about our senses, related body parts, and how they help us learn about the world around us.

Connecting Learning

1. What are the five senses? [sight, hearing, taste, touch, smell]
2. What do we use our nose for? [smelling]
3. What body part do we use to hear the sounds around us? [ears]
4. Could you see without your mouth? Why or why not? [Yes. We use our mouths to taste and our eyes to see.]
5. What body part do you use to touch things? [hands, usually] Can you use other body parts? Explain. [Yes. Most places on our body are sensitive to touch.]
6. Which of our fives senses do you think is most important? Explain.

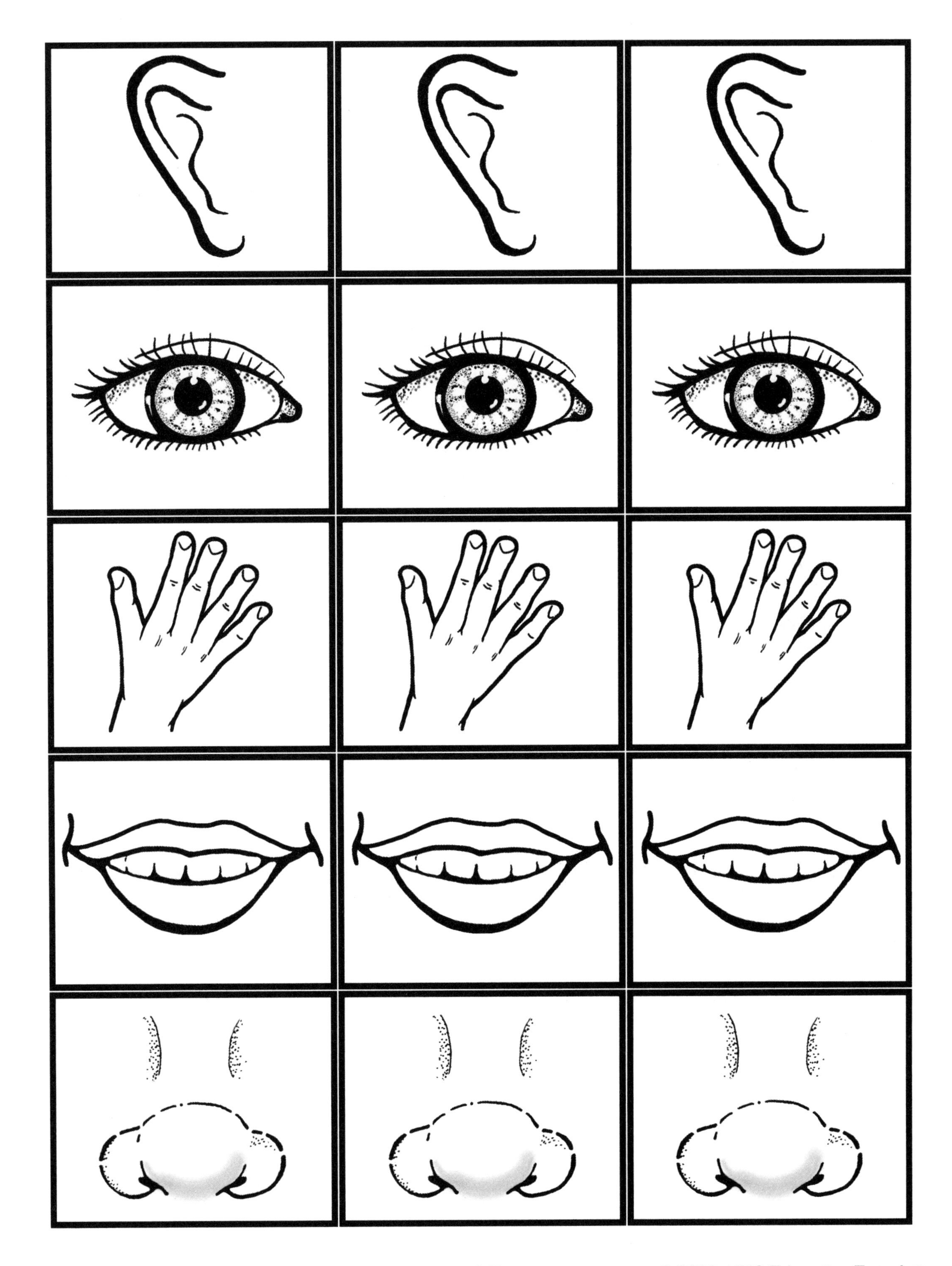

Taste	**Taste**	**Taste**
Touch	**Touch**	**Touch**
Sight	**Sight**	**Sight**
Smell	**Smell**	**Smell**
Hearing	**Hearing**	**Hearing**

Bibliography

Teacher Resources

Ardley, Neil. *The Science Book of Color.* Gulliver Books. Harcourt Brace Jovanovich. New York. 1991.

Silverstein, Alvin, Virginia and Robert. *Smell, The Subtle Sense.* Morrow Junior Books. New York. 1992.

Taylor, Barbara. *Seeing Is Not Believing!* Random House. New York. 1990.

The Sense of Sight

Carle, Eric. *Brown Bear, Brown Bear, What Do You See?* Henry Holt. New York. 1992.

Ehlert, Lois. *Planting A Rainbow.* Harcourt Brace Jovanovich. New York. 1988.

Freeman, Don. *A Rainbow Of My Own.* Viking Press. New York. 1966.

Martin, Bill Jr. and John Archambault. *Knots on a Counting Rope.* Henry Holt and Company. New York. 1997.

Micklethwait, Lucy. *I Spy: Shapes In Art.* HarperCollins. New York. 2004.

Micklethwait, Lucy. *I Spy: An Alphabet In Art.* Greenwillow Books. New York. 1996.

Rosen, Michael. *We're Going On a Bear Hunt.* McElderry. New York. 2003.

Stinson, Kathy. *Red Is Best.* Annick Press. Toronto, Canada. 1992.

Walsh, Ellen Stoll. *Mouse Paint.* Voyager Books. San Diego, CA. 1989.

Williams, Sue. *I Went Walking.* Voyager Books. San Diego, CA. 1989.

The Sense of Touch

Aliki. *My Hands.* HarperCollins. New York. 1990.

Blackstein, Karen. *The Blind Men and the Elephant.* Scholastic, Inc. New York. 1992.

Dorros, Arthur. *Feel the Wind.* HarperCollins. New York. 1989.

The Sense of Taste

Carle, Eric. *The Very Hungry Caterpillar.* Philomel Books. New York. 1987.

Dooley, Norah. *Everybody Cooks Rice.* Carolrhoda Books. Minneapolis, MN. 1991.

Marshall, James. *Goldilocks and The Three Bears.* Dial Books. New York. 1988.

Seuss, Dr. *Green Eggs and Ham.* Random House. New York. 1960.

Seuss, Dr. *Scrambled Eggs Super!* Random House. New York. 1953.

Sharmat, Mitchell. *Gregory The Terrible Eater.* Scholastic, Inc. New York. 1980.

Tolhurst, Marilyn. *Somebody and the Three Blairs.* Orchard Books. New York. 1990.

Turkle, Brinton. *Deep In The Forest.* Dutton Children's Books. New York. 1976.

The Sense of Smell

Howard, Katherine. *Little Bunny Follows His Nose.* Golden Books. New York. 1971.

Krauss, Ruth. *The Happy Day.* HarperCollins. New York. 1949.

The Sense of Hearing

Baylor, Byrd. *The Other Way To Listen.* Aladdin Paperbacks. New York. 1997.

Carle, Eric. *The Very Quiet Cricket.* Philomel Books. New York. 1990.

Fox, Mem. *Night Noises.* Voyager Books New York. 1992.

Grindley, Sally. *Shh!* Hodder Children's Books. London. 2006.

Martin, Bill, Jr. and Eric Carle. *Polar Bear, Polar Bear, What Do You Hear?* Henry Holt and Company. New York. 1991.

Martin, Bill, Jr. and John Archambault. *Listen To the Rain.* Henry Holt and Company. New York. 1988.

Perkins, Al. *The Ear Book.* Random House. New York. 2007.

Peterson, Jeanne Whitehouse. *I Have A Sister My Sister Is Deaf.* HarperCollins. New York. 1984.

Showers, Paul. *The Listening Walk.* HarperTrophy. New York. 1993.

Multi-Sensory
Aliki. *My Five Senses.* HarperCollins. New York. 1989.

Cole, Joanna and Bruce Degen. *The Magic School Bus Explores the Senses.* Scholastic, Inc. New York. 1999.

Glander, Lisa Jayne. *How Do You Know? A Book About the Five Senses.* Tate Publishing. Mustang, OK. 2007.

Heide, Forence Parry and Judith Heide Gilliand. *The Day of Ahmed's Secret.* Mulberry Books. New York. 1995.

Hutchins, Pat. *Changes, Changes.* Aladdin Paperbacks. New York. 1987.

Martin, Bill and John Archambault. *Here Are My Hands.* Henry Holt and Company. New York. 2007.

Ziefert, Harriet. *You Can't Taste a Pickle With Your Ear: A Book About Your 5 Senses.* Blue Apple Books. Maplewood, NJ. 2002.

The AIMS Program

AIMS is the acronym for "**A**ctivities **I**ntegrating **M**athematics and **S**cience." Such integration enriches learning and makes it meaningful and holistic. AIMS began as a project of Fresno Pacific University to integrate the study of mathematics and science in grades K-9, but has since expanded to include language arts, social studies, and other disciplines.

AIMS is a continuing program of the non-profit AIMS Education Foundation. It had its inception in a National Science Foundation funded program whose purpose was to explore the effectiveness of integrating mathematics and science. The project directors, in cooperation with 80 elementary classroom teachers, devoted two years to a thorough field-testing of the results and implications of integration.

The approach met with such positive results that the decision was made to launch a program to create instructional materials incorporating this concept. Despite the fact that thoughtful educators have long recommended an integrative approach, very little appropriate material was available in 1981 when the project began. A series of writing projects ensued, and today the AIMS Education Foundation is committed to continuing the creation of new integrated activities on a permanent basis.

The AIMS program is funded through the sale of books, products, and professional-development workshops, and through proceeds from the Foundation's endowment. All net income from programs and products flows into a trust fund administered by the AIMS Education Foundation. Use of these funds is restricted to support of research, development, and publication of new materials. Writers donate all their rights to the Foundation to support its ongoing program. No royalties are paid to the writers.

The rationale for integration lies in the fact that science, mathematics, language arts, social studies, etc., are integrally interwoven in the real world, from which it follows that they should be similarly treated in the classroom where students are being prepared to live in that world. Teachers who use the AIMS program give enthusiastic endorsement to the effectiveness of this approach.

Science encompasses the art of questioning, investigating, hypothesizing, discovering, and communicating. Mathematics is a language that provides clarity, objectivity, and understanding. The language arts provide us with powerful tools of communication. Many of the major contemporary societal issues stem from advancements in science and must be studied in the context of the social sciences. Therefore, it is timely that all of us take seriously a more holistic method of educating our students. This goal motivates all who are associated with the AIMS Program. We invite you to join us in this effort.

Meaningful integration of knowledge is a major recommendation coming from the nation's professional science and mathematics associations. The American Association for the Advancement of Science in *Science for All Americans* strongly recommends the integration of mathematics, science, and technology. The National Council of Teachers of Mathematics places strong emphasis on applications of mathematics found in science investigations. AIMS is fully aligned with these recommendations.

Extensive field testing of AIMS investigations confirms these beneficial results:

1. Mathematics becomes more meaningful, hence more useful, when it is applied to situations that interest students.
2. The extent to which science is studied and understood is increased when mathematics and science are integrated.
3. There is improved quality of learning and retention, supporting the thesis that learning which is meaningful and relevant is more effective.
4. Motivation and involvement are increased dramatically as students investigate real-world situations and participate actively in the process.

We invite you to become part of this classroom teacher movement by using an integrated approach to learning and sharing any suggestions you may have. The AIMS Program welcomes you!

AIMS Education Foundation Programs

When you host an AIMS workshop for elementary and middle school educators, you will know your teachers are receiving effective, usable training they can apply in their classrooms immediately.

AIMS Workshops are Designed for Teachers

- Correlated to your state standards;
- Address key topic areas, including math content, science content, and process skills;
- Provide practice of activity-based teaching;
- Address classroom management issues and higher-order thinking skills;
- Give you AIMS resources; and
- Offer optional college (graduate-level) credits for many courses.

AIMS Workshops Fit District/Administrative Needs

- Flexible scheduling and grade-span options;
- Customized (one-, two-, or three-day) workshops meet specific schedule, topic, state standards, and grade-span needs;
- Prepackaged four-day workshops for in-depth math and science training available (includes all materials and expenses);
- Sustained staff development is available for which workshops can be scheduled throughout the school year;
- Eligible for funding under the Title I and Title II sections of No Child Left Behind; and
- Affordable professional development—consecutive-day workshops offer considerable savings.

University Credit—Correspondence Courses

AIMS offers correspondence courses through a partnership with Fresno Pacific University.

- Convenient distance-learning courses—you study at your own pace and schedule. No computer or Internet access required!

Introducing AIMS State-Specific Science Curriculum

Developed to meet 100% of your state's standards, AIMS' State-Specific Science Curriculum gives students the opportunity to build content knowledge, thinking skills, and fundamental science processes.

- Each grade-specific module has been developed to extend the AIMS approach to full-year science programs. Modules can be used as a complete curriculum or as a supplement to existing materials.
- Each standards-based module includes math, reading, hands-on investigations, and assessments.

Like all AIMS resources, these modules are able to serve students at all stages of readiness, making these a great value across the grades served in your school.

For current information regarding the programs described above, please complete the following form and mail it to: P.O. Box 8120, Fresno, CA 93747.

Information Request

Please send current information on the items checked:

____ *Basic Information Packet* on AIMS materials

____ Hosting information for AIMS workshops

____ AIMS State-Specific Science Curriculum

Name: ____________________________________

Phone: __________________ E-mail: __________________

Address: ____________________________________

Street City State Zip

YOUR K-9 MATH AND SCIENCE
CLASSROOM ACTIVITIES RESOURCE

The AIMS Magazine is your source for standards-based, hands-on math and science investigations. Each issue is filled with teacher-friendly, ready-to-use activities that engage students in meaningful learning.

- *Four issues each year (fall, winter, spring, and summer).*

Current issue is shipped with all past issues within that volume.

1824	Volume XXIV	2009-2010	$19.95
1825	Volume XXV	2010-2011	$19.95

Two-Volume Combination

M20810	Volumes XXIII & XXIV	2008-2010	$34.95
M20911	Volumes XXIV & XXV	2009-2011	$34.95

Complete volumes available for purchase:

1802	Volume II	1987-1988	$19.95
1804	Volume IV	1989-1990	$19.95
1805	Volume V	1990-1991	$19.95
1807	Volume VII	1992-1993	$19.95
1808	Volume VIII	1993-1994	$19.95
1809	Volume IX	1994-1995	$19.95
1810	Volume X	1995-1996	$19.95
1811	Volume XI	1996-1997	$19.95
1812	Volume XII	1997-1998	$19.95
1813	Volume XIII	1998-1999	$19.95
1814	Volume XIV	1999-2000	$19.95
1815	Volume XV	2000-2001	$19.95
1816	Volume XVI	2001-2002	$19.95
1817	Volume XVII	2002-2003	$19.95
1818	Volume XVIII	2003-2004	$19.95
1819	Volume XIX	2004-2005	$19.95
1820	Volume XX	2005-2006	$19.95
1821	Volume XXI	2006-2007	$19.95
1822	Volume XXII	2007-2008	$19.95
1823	Volume XXIII	2008-2009	$19.95

Volumes II to XIX include 10 issues.

Call 1.888.733.2467 or go to www.aimsedu.org

Subscribe to the AIMS Magazine

$19.95 a year!

AIMS Magazine is published four times a year.

Subscriptions ordered at any time will receive all the issues for that year.

To see all that AIMS has to offer, check us out on the Internet at www.aimsedu.org. At our website you can preview and purchase AIMS books and individual activities, learn about State-Specific Science and Essential Math, explore professional development workshops and online learning opportunities, search our activities database, buy manipulatives and other classroom resources, and download free resources including articles, puzzles, and sample AIMS activities.

AIMS E-mail Specials
While visiting the AIMS website, sign up for our FREE e-mail newsletter with monthly subscriber-only specials. You'll also receive advance notice of new products.

Sign up today!

AIMS Program Publications

Actions With Fractions, 4-9
The Amazing Circle, 4-9
Awesome Addition and Super Subtraction, 2-3
Bats Incredible! 2-4
Brick Layers II, 4-9
The Budding Botanist, 3-6
Chemistry Matters, 4-7
Counting on Coins, K-2
Cycles of Knowing and Growing, 1-3
Crazy About Cotton, 3-7
Critters, 2-5
Earth Book, 6-9
Electrical Connections, 4-9
Exploring Environments, K-6
Fabulous Fractions, 3-6
Fall Into Math and Science*, K-1
Field Detectives, 3-6
Finding Your Bearings, 4-9
Floaters and Sinkers, 5-9
From Head to Toe, 5-9
Glide Into Winter With Math and Science*, K-1
Gravity Rules! 5-12
Hardhatting in a Geo-World, 3-5
Historical Connections in Mathematics, Vol. I, 5-9
Historical Connections in Mathematics, Vol. II, 5-9
Historical Connections in Mathematics, Vol. III, 5-9
It's About Time, K-2
It Must Be A Bird, Pre-K-2
Jaw Breakers and Heart Thumpers, 3-5
Looking at Geometry, 6-9
Looking at Lines, 6-9
Machine Shop, 5-9
Magnificent Microworld Adventures, 6-9
Marvelous Multiplication and Dazzling Division, 4-5
Math + Science, A Solution, 5-9
Mathematicians are People, Too
Mathematicians are People, Too, Vol. II
Mostly Magnets, 3-6
Movie Math Mania, 6-9
Multiplication the Algebra Way, 6-8
Out of This World, 4-8
Paper Square Geometry:
  The Mathematics of Origami, 5-12
Puzzle Play, 4-8
Pieces and Patterns*, 5-9
Popping With Power, 3-5
Positive vs. Negative, 6-9
Primarily Bears*, K-6
Primarily Earth, K-3
Primarily Magnets, K-2
Primarily Physics*, K-3
Primarily Plants, K-3
Primarily Weather, K-3
Problem Solving: Just for the Fun of It! 4-9
Problem Solving: Just for the Fun of It! Book Two, 4-9
Proportional Reasoning, 6-9
Ray's Reflections, 4-8
Sensational Springtime, K-2
Sense-Able Science, K-1
Shapes, Solids, and More: Concepts in Geometry, 2-3
The Sky's the Limit, 5-9
Soap Films and Bubbles, 4-9
Solve It! K-1: Problem-Solving Strategies, K-1
Solve It! 2nd: Problem-Solving Strategies, 2
Solve It! 3rd: Problem-Solving Strategies, 3
Solve It! 4th: Problem-Solving Strategies, 4
Solve It! 5th: Problem-Solving Strategies, 5
Solving Equations: A Conceptual Approach, 6-9
Spatial Visualization, 4-9
Spills and Ripples, 5-12
Spring Into Math and Science*, K-1
Statistics and Probability, 6-9
Through the Eyes of the Explorers, 5-9
Under Construction, K-2
Water, Precious Water, 4-6
Weather Sense: Temperature, Air Pressure, and Wind, 4-5
Weather Sense: Moisture, 4-5
What's Next, Volume 1, 4-12
What's Next, Volume 2, 4-12
What's Next, Volume 3, 4-12
Winter Wonders, K-2

Essential Math

Area Formulas for Parallelograms, Triangles, and Trapezoids, 6-8
Measurement of Prisms, Pyramids, Cylinders, and Cones, 6-8
Circumference and Area of Circles, 5-7
Measurement of Rectangular Solids, 5-7
Perimeter and Area of Rectangles, 4-6
The Pythagorean Relationship, 6-8

Spanish Edition

Constructores II: Ingeniería Creativa Con Construcciones LEGO®, 4-9
The entire book is written in Spanish. English pages not included.

* Spanish supplements are available for these books. They are only available as downloads from the AIMS website. The supplements contain only the student pages in Spanish; you will need the English version of the book for the teacher's text.

For further information, contact:
AIMS Education Foundation • P.O. Box 8120 • Fresno, California 93747-8120
www.aimsedu.org • 559.255.6396 (fax) • 888.733.2467 (toll free)

Duplication Rights

No part of any AIMS books, magazines, activities, or content—digital or otherwise—may be reproduced or transmitted in any form or by any means—including photocopying, taping, or information storage/retrieval systems—except as noted below.

Standard Duplication Rights

- A person or school purchasing AIMS activities (in books, magazines, or in digital form) is hereby granted permission to make up to 200 copies of any portion of those activities, provided these copies will be used for educational purposes and only at one school site.
- Workshop or conference presenters may make one copy of any portion of a purchased activity for each participant, with a limit of five activities per workshop or conference session.
- All copies must bear the AIMS Education Foundation copyright information.

Standard duplication rights apply to activities received at workshops, free sample activities provided by AIMS, and activities received by conference participants.

Unlimited Duplication Rights

Unlimited duplication rights may be purchased in cases where AIMS users wish to:
- make more than 200 copies of a book/magazine/activity,
- use a book/magazine/activity at more than one school site, or
- make an activity available on the Internet (see below).

These rights permit unlimited duplication of purchased books, magazines, and/or activities (including revisions) for use at a given school site.

Activities received at workshops are eligible for upgrade from standard to unlimited duplication rights.

Free sample activities and activities received as a conference participant are not eligible for upgrade from standard to unlimited duplication rights.

State-Specific Science modules are licensed to one classroom/one teacher and are therefore not eligible for upgrade from standard to unlimited duplication rights.

Upgrade Fees

The fees for upgrading from standard to unlimited duplication rights are:
- $5 per activity per site,
- $25 per book per site, and
- $10 per magazine issue per site.

The cost of upgrading is shown in the following examples:
- activity: 5 activities x 5 sites x $5 = $125
- book: 10 books x 5 sites x $25 = $1250
- magazine issue: 1 issue x 5 sites x $10 = $50

Purchasing Unlimited Duplication Rights

To purchase unlimited duplication rights, please provide us the following:
1. The name of the individual responsible for coordinating the purchase of duplication rights.
2. The title of each book, activity, and magazine issue to be covered.
3. The number of school sites and name of each site for which rights are being purchased.
4. Payment (check, purchase order, credit card)

Requested duplication rights are automatically authorized with payment. The individual responsible for coordinating the purchase of duplication rights will be sent a certificate verifying the purchase.

Internet Use

AIMS materials may be made available on the Internet if all of the following stipulations are met:
1. The materials to be put online are purchased as PDF files from AIMS (i.e., no scanned copies).
2. Unlimited duplication rights are purchased for all materials to be put online for each school at which they will be used. (See above.)
3. The materials are made available via a secure, password-protected system that can only be accessed by employees at schools for which duplication rights have been purchased.

AIMS materials may not be made available on any publicly accessible Internet site.